Praise for *Th* *from th*

"In this lovely, thoughtful series of meditations on what the author calls the 'simple yet sacred' moments of our lives, Charlene Costanzo calls upon the beauty of the natural world outside ourselves as well as the quiet wisdom that lies within us, offering strength and healing. This book is a gift of peace and comfort for all readers."

—**Reeve Lindbergh**, author of multiple books
including *Two Lives* and *Under a Wing*

"*The Twelve Gifts from the Garden* is a joyful read bursting with the beauty of nature and reflections of lessons learned in life. I have been a fan of Charlene Costanzo's daily emails and books for many years. Most mornings, reading her reflections gives me goosebumps because they are like having a cup of coffee with God. They are what I would imagine God would be saying to me in that moment because Charlene's reflections would be in line with what I prayed the night before. This is a wonderful book to gift yourself and others!"

—**Gina La Benz**, author of *Anchor Moments: Hope,*
Healing, and Forgiveness

"From the very first paragraph, I felt my spirit stirred—like when you don't know how thirsty you are until that first sip of cool water. Instantly, I was transported to an enchanted island to experience insightful gifts alongside the author. Her prose is as soothing as the scenes she describes. Charlene will renew your wonder and gratitude for our beautiful earth, and for the gifts and lessons available to us all when we're mindfully watching."

—**Angela Howell**, author of *Finding the Gift: Daily Meditations for Mindfulness*

"After reading *The Twelve Gifts from the Garden*, I may never pass a plant again without feeling a nudge toward reflection and introspection. In this book, Charlene gently reminds us to live more mindfully, to be present in the moment, to not worry about our harvest but to focus on how we plant our seeds. She does this by sharing lessons prompted by experiences in a beautifully described garden. Throughout the book I felt myself commiserate, empathize, and celebrate with her in her disappointments, struggles, and triumphs. Charlene skillfully comforts and soothes the reader, while at the same time challenging us to think differently and learn more from nature, particularly about how to use the innate qualities she calls the Twelve Gifts. I didn't want the book to end, and I know I will read it again and again."

—**Adrienne Falzon,** author of *The Search for the Perfect Shell, What is an Angel?, Selfish Sally, It's Not Fair!,* and *Live Like Paul*

"Just as the waves of the sea ebb and flow, so do the currents of our lives. Charlene Costanzo's *The Twelve Gifts from the Garden* with its stunning imagery and poignant recollections, reminds us to savor the pleasures and discoveries of the day, the healing power of nature, and the joy of connection—to oneself and others."

> —**Maryann Ridini Spencer**, award-winning screenwriter and bestselling author of *Lady in the Window* and *The Paradise Table*

"With the wonder of a child, the introspection of a mature woman, and the wisdom of the elders, Charlene Costanzo inspires us to discover the hidden treasures of nature as we make our way through this journey called life. In *The Twelve Gifts from the Garden*, her accounts of the blessings that she has gleaned from her journey move us to wonder. They arouse joyful anticipation of the blessings awaiting us on our own journeys. It is said that there are those who live in a constant state of amazement. This clearly describes Charlene Costanzo and those who follow her lead."

> —**Gloria Gaynor**, singer, Grammy Award-winner for "I Will Survive" and "Testimony" and author of *I Will Survive: The Book* and *We Will Survive: True Stories of Encouragement, Inspiration, and the Power of Song*

"*The Twelve Gifts from the Garden* invites us to see the beauty of Sanibel through the imaginative perspective of its author. While taking readers on a journey through lush tropical greenery, Charlene Costanzo offers parables of life and love that will last a lifetime. This is a lovely gift to give to oneself and to others."

—**Charles Sobezak**, author of *Living Sanibel: A Nature Guide to Sanibel* and *Captiva Islands*, *The Living Gulf Coast*, and *Six Mornings on Sanibel*

"This book is a feast, a dream, a wondrous adventure. Gift yourself with this exquisite collection of stories by Charlene Costanzo and you will feel transported from start to finish. Charlene is our gentle guide through inner and outer landscapes filled with meandering paths, curious questions, and colorful surprises. Along the way, you'll meet black butterflies, beautyberry plants, pom pom palms—and, most importantly, your own inner garden. Our lives are fleeting gifts; Costanzo reminds us of this with every perfectly crafted sentence along this enchanting journey of a book."

—**Sherry Richert Belul**, founder of Simply Celebrate and author of *Say It Now: 33 Creative Ways to Say I Love You to the Most Important People in Your Life*

The Twelve Gifts
from
the Garden

The Twelve Gifts from the Garden

Life Lessons for Peace and Well-Being

by Charlene Costanzo

CORAL GABLES

Cover, Layout & Design: Morgane Leoni
Cover illustration: © moleskostudio & © biscotto87

For permission requests, please contact the publisher at:
Mango Publishing Group
2850 S Douglas Road, 2nd Floor
Coral Gables, FL 33134 USA
info@mango.bz

For special orders, quantity sales, course adoptions and corporate sales, please
email the publisher at sales@mango.bz. For trade and wholesale sales, please
contact Ingram Publisher Services at customer.service@ingramcontent.com or
+1.800.509.4887.

The Twelve Gifts from the Garden: Life Lessons for Peace and Well-Being

Library of Congress Cataloging-in-Publication number: 2020940962
ISBN: (print) 978-1-64250-372-2, (ebook) 978-1-64250-373-9
BISAC category code OCC019000, BODY, MIND & SPIRIT / Inspiration &
Personal Growth

Printed in the United States of America

With love and gratitude
for my mother and grandmothers
and Mother Earth

Contents

Note to the Reader

Readers are asked to imagine walking in the author's sandals—or at least by her side—and perceiving the flowers, roots, leaves, and trees as she did, which was sometimes with whimsy, often with wonder, always with awe. She hopes that *The Twelve Gifts from the Garden* will give a measure of enrichment and empowerment to readers as they journey through life in their own shoes.

Preface

The Story behind the Twelve Gifts

One morning in 1987, when my daughters were teenagers nearing high school graduation, I woke with a shock. My children were about to leave home, the most critical years of their development were over, and I was just beginning to understand that unconditional love was the most important thing I could give them. With hindsight, I wished I had done some things differently.

Weeks later, once again I woke with strong emotion. This time it was a feeling of euphoria. In the sleep state, I had been in a place where I heard about twelve gifts. I remembered a few of them: Strength, Courage, Beauty, Compassion, Joy... and I recalled a repeated phrase: "May you... May you... May you..." with what felt like a bestowing of blessings. As I moved into wakefulness, the details of the dream evaporated. Holding onto wisps of it, I wrote a message and fashioned it into a booklet titled *Welcome to the World: The Twelve Gifts of Birth*. It was what I wished I had whispered in my babies' ears and said often as they grew. It told them they were born with gifts. Gentle wishes suggested how to use each gift to live well. I regretted that I had not articulated the message earlier and used it to guide my daughters. But I was just beginning to comprehend it myself.

I felt strongly that all children, not just my own, deserve to hear that they are worthy and gifted. I wanted to see *The Twelve Gifts of Birth* published. With high hopes, I prepared submission packages. After the twentieth rejection, I decided to give up on selling *The Twelve Gifts of Birth* to an established publisher. I resolved to publish it myself—someday. For years "someday" was a vague, elusive time in the future. But every once in a while, I'd feel a push forward. In 1995, I became increasingly disturbed by news stories of abused children. I realized that the message of *The Twelve Gifts of Birth* held potential to help in a small way, and my resolve strengthened. But still I said, *someday*.

A year later, sitting quietly in my mom's hospital room one afternoon, a month before she died, I heard a voice within me declare *What you do with your time and talent is critically important. Pay attention*. I knew immediately what the admonition to pay attention meant. It was time to embrace "someday" and act upon what was calling me—my book.

During the entire time I was preparing *The Twelve Gifts of Birth* for publication, I experienced firsthand that miracles do happen when we follow our bliss in the spirit of service. When work is a labor of love, doors open. That was a premise and a promise that I had heard from many sources. But, although I believed it, never before had I acted as if it were true.

It took a year and half of full-time work and a substantial amount of money—more than I had saved for this purpose—to bring the richly illustrated gift book into reality. In unexpected and surprising ways, all the resources that I needed along the way, financial and otherwise, appeared in perfect time. There were times when the steps I took seemed wrong or unnecessary, but later I saw how each "false" step became a stepping-stone.

The Twelve Gifts of Birth was released in September 1998. I had anticipated that the book would do well, but the market's response surpassed my expectations. Some readers shared how the book affected them. And they weren't just parents of young children. The first letter I received said, "I am seventy-two years old and have spent years of my life in therapy. I grew up believing that I was worthy only if I accomplished my goals and made a lot of money. My mind and heart have been healed by these very twelve gifts. I realize that I live by them today, but we both know they have been mine all along."

The story is a message intended for all the children of the world, children of all ages, colors, creeds, and cultures. It begins during a time when royal gifts were pronounced by wise godmothers upon princes and princesses at their birth.

The gifts were intangible virtues, resources, and qualities that enrich one's sense of self-worth and dignity and enhance one's ability to make a difference in the world.

Eventually, it becomes clear to the wise godmothers that the gifts are not only intended for all children, but are actually inherent in all children. The godmothers yearned to make this known to everyone. But announcing the gifts to all was not allowed in the kingdom at that time.

The godmothers predict, however, that, someday, all the world's children will learn of their noble inheritance and birthright gifts. When that happens, "a miracle will unfold on the kingdom of Earth," they say.

Readers then learn of their strength, beauty, courage, compassion, hope, joy, talent, imagination, reverence, wisdom, love, and faith, and receive a guiding wish to use each gift well.

In the process of writing and publishing *The Twelve Gifts of Birth*, I realized that I needed to hear its message myself. Repeatedly. And I needed to understand the gifts better. Since then, they've never been far from my mind. Don't get me wrong; I don't live in a perpetually blissful state of awareness and love. Of course at times I get upset, angry, petty, judgmental... But I am continually trying to better recognize and cultivate the gifts in myself and to see them and kindle them in others. I've been working on this through, among other things, mindfulness, prayer, reading, writing, workshops, and study.

I especially appreciate my studies in philosophy and spiritual psychology. Beyond the degrees I received, I value how education stretched me, filled me, and in its way also emptied me, so I could be open to receive and better evaluate thoughts and ideas. I hold education in high regard. Just as my heart aches for all children to know that they are gifted, talented and valuable, I yearn for high-quality education to be available to all children. All people.

Of course there's the great school of life. We're all in it. With or without formal education, it's living that gives the biggest, the smallest, the most basic, and the most critical lessons to every one of us. We all get our share, like it or not.

I don't get all of my lessons right. I've experienced a lot of "aha!" moments, but also a good number of "uh-ohs." I'm not a straight A student in the school of life. Sometimes it feels like I'm repeating a lesson, or even a whole grade. But through each loss, gain, challenge, and triumph, I've grown. I'm still growing. And learning.

Life has invited me, and continues to encourage me, to walk my talk. I'm grateful for every nudge. Some are more like pushes. But in addition to teaching through tough lessons and challenges, life gives answers through joy, imagination, and beauty. And nature. Sometimes it seems to me as if

Mother Nature is there right in front of me, trying to hand out answers on silver platters. Or green leaves, brown trunks, or purple flowers. Learning does not have to be a struggle; it also comes with grace and ease, comfort and peace, and fun. We just need to be open to learning from unexpected places and situations.

I'm as passionate about understanding and living the Twelve Gifts as I was when I woke from that dream in 1987. In trying to consciously use the Twelve Gifts in responding to life's ups and downs, I've ended up writing a lot more about them—books, blogs, and daily email messages. *The Twelve Gifts from the Garden* emerged out of my time spent in nature.

Introduction

Dear Reader,

I am not a master gardener. However, I have an abundance of appreciation for all things that sprout, grow, blossom, and bloom. I'm grateful for how plants soothe us and uplift us. I'm thankful that they feed our bodies, enrich our minds, and nourish our souls. Plants help us breathe. They have healing power. Wordlessly, they lead us toward understanding. They teach by example.

I've received a lot of guidance from plants, including lessons related to strength, beauty, courage, compassion, hope, joy, talent, imagination, reverence, wisdom, love, and faith. This is what I have to share, what I wish to share, in this book. If you are already familiar with my work, you know that I'm passionate about these twelve resources—which I call the Twelve Gifts. If you are not yet familiar with the Twelve Gifts, I hope you soon will be, by starting here. Familiar or not, I'd like to tell you what's in this book and why I wrote it.

The Twelve Gifts from the Garden is a collection of discoveries, healing perceptions, and aha experiences I've had on Sanibel, an island off the southwest coast of Florida.

Most events were triggered in a garden or in nature. Usually they were stirred by a "close encounter" with a plant. Each sharing contains something, often a lesson, about using our twelve inner gifts.

I was inspired, in part, to collect my musings and publish these "gifts from the garden" because I thoroughly enjoyed reading Anne Morrow Lindbergh's *Gift from the Sea*, which she wrote in the early 1950s on Florida's Captiva Island. I have appreciated her thinking, admired her style, and delighted in the role seashells play in her essays. Although I was just a toddler at the time she wrote the bestseller, and I did not discover *Gift from the Sea* until I was almost forty, I have taken my own lessons from nature, especially plants, since my early childhood. And what Anne Morrow Lindberg did with shells on Captiva, I started doing with plants upon my first visit to Sanibel Island, well before I discovered her wonderful book. Please don't compare my writing with her exquisite essays. Let the reflections in both books stand on their own. If you have not yet read *Gift from the Sea*, I highly recommend it. Right now, I'd like to shed light on how I began taking lessons from nature. As you read my story, consider your own relationship with flora and fauna.

Until I was ten years old, my parents and I lived in a redbrick ten-family apartment house in Linden, New Jersey. Perhaps because our apartment building was almost entirely surrounded by concrete, I found comfort in a small, neglected patch of dirt adjacent to our building. It served as my first garden. Enclosed within an unpainted picket fence, that desolate space sprung to life each summer when grasses tipped with tiny purple, orange, and yellow flowers filled the area. Though others called them weeds, those grasses stirred my joy and taught me that good things can be present in unpleasant circumstances.

On the opposite side from our apartment building stood a two-story home that housed a neighborhood tavern. For a time, I disliked that drab gray building. A large tavern sign hung above the porch steps. Beer advertisements glowed in the windows. Rheingold. Pabst. Schlitz. Through my eyes, the neon-decorated building seemed out of place among family homes.

But on midsummer mornings, when I looked through our kitchen window, my heart opened with gratitude and joy. From that window I saw no tavern, just masses of morning glories blooming bright and blue against the weathered gray clapboard on the side of that house. The flowers looked so alive, so pure. The vibrant sight of them climbing a large trellis thrilled me. I loved them to tears. That taught me that a shift in perspective can transform an experience.

One day, noticing colorful clothes of varied sizes hanging on clotheslines behind the tavern house, I realized that a family like mine lived there. My opinion of the building softened further. My initial observations, judgments, and feelings about that neighboring house played a part in my learning to look beyond first impressions and to see situations from different vantage points.

From wildflowers pushing through nearby sidewalk cracks I concluded that life has strength and determination. And if plants can thrive in unfavorable conditions, I can too. Since my first "garden," I've been drawn to all sorts of green places. City parks. Cemeteries. Nurseries. Nature preserves. Each has taught me something.

In this book, I sometimes reach back to earlier times and other places I've experienced. I also jump around in time. Mostly, I share the insights I gained in a botanical garden on Sanibel Island. As you read, imagine you are walking the paths with me. Or envision being in the garden on your own or with loved ones. Notice what resonates within you. Listen and feel for your own insights. Even when you're not physically present, this garden holds gifts for you. All of nature does.

Wishing you the best of life's gifts,

—CHARLENE

First Crossing

When you cross a bridge,
you take a break from this world!

—Mehmet Murat ildan

The first time I saw San Carlos Bay and glimpsed our destination on the other end of the causeway, joy surged faster than my heart could swell to contain it. My chest ached with an emotion I couldn't name. I immediately felt deeply connected to this particular place.

Setting aside the AAA TripTik map that highlighted our route from Jamestown, New York, to Sanibel, Florida, I gaped at the expanse of turquoise water. My husband, our two young daughters, and I were initially silent, awed by what we saw.

Golden ripples reflecting the late afternoon sun twinkled like daytime stars to wish upon. Waves made by boaters intersected and formed mesmerizing patterns in shades of aquamarine. I heard "Wow" a few times. I said it myself as we crossed the three bridges and two tiny islands that make up the causeway. We cranked open our windows, all at the same time as fast as we could. A warm, sea-scented breeze cleared the car air and twirled our hair. A little more speed brought hot wind and laughter. On the final bridge, a group of pelicans roosted on the rail while a row of them glided single file above, like a welcoming committee.

On the shore to the right of the bridge, several bayfront houses stood well apart from one another. To the left of the bridge, there were a few low condominiums. Other than that, all we saw was green. Nothing man-made rose above the trees and foliage. The island seemed to exude a vibrant, humming strength. A peaceful strength of harmony and balance. It's no wonder: nearly half the island is a protected wildlife preserve. Natural.

As the car passed off the bridge onto land, I felt as if I were coming home to a place I'd never been before. Although I fell in love with the island even before our car reached its shore, I didn't imagine I'd be returning many times. Nor did I sense the lessons and gifts it had in store for me.

The Pom-Pom Palm

Find ecstasy in life;
the mere sense of living is joy enough.

—Emily Dickinson

In the world of plants, just as with people, we get first impressions of personalities and character. At least I do. Some trees in this garden give me an impression of lightness and playfulness. To me they are reminders to have fun and find the funny things in life as much as possible. They stir my joy too.

The pom-pom palm is one of them. That's just my name for it. Its most common name is ponytail palm. Officially named *Beaucarnea recurvata*, this tree is also called the elephant foot tree. Very large and thick at the bottom, with grayish bark resembling wrinkled skin, the trunk does look like an elephant's foot. But I think the name pom-pom palm is a good fit. The leafy parts of the top, the long skinny frond leaves, stick out in bunches, looking messy like a teenager's bedhead. If there was just one bunch, I'd stick with the name ponytail palm, and if there were two I might call it a pigtail tree. But with several, this particular one reminds me of a cheerleader's pom-poms waving in the wind. This tree's got spirit.

It's one of the unofficial greeters to anyone coming to tour the garden or to check in at the registration office at Sanibel Moorings, the resort on the island whose grounds are a certified botanical garden. It stands here giving a warm welcome to every passerby too. Or maybe to excited children arriving on vacation for the first time, it chants an upbeat cheer, "Hey, hey, are you ready to play?!"

I imagine it as able to read the emotions of humans, too, so that it knows just how much cheer to extend; it's a sensitive tree. Sometimes it just looks cute and caring enough to make a person smile. I know it always lifts me up a little higher than where I was before I saw it. This tree is a model to me, a reminder to give everyone I meet a bit of cheer, in whatever form they may need.

Am I anthropomorphizing? Maybe. Using my imagination? Yes, of course. But there's more to this. We don't yet know what it's like to be a tree. Of course we likely never will, not fully. One can never really know what it's like to be in someone else's shoes. Or roots in this case. But scientists are learning that trees have intelligence. They help each other heal. They communicate with other trees through soil and air by releasing certain chemicals, and they even warn each other of danger. Really.

German forester and author Peter Wohlleben says that trees are sentient beings. They feed their young, they cooperate, they communicate, and they have character too. There is just so much more to every living thing that we don't yet know. It's exciting to think of what else scientists will learn about the inner lives of plants.

In a smaller form than this one, the pom-pom palm, a.k.a. ponytail palm, a.k.a. elephant foot tree, is commonly a house plant. It seems like a fun tree to remind us to be of good cheer. I think it might just be the next addition to my household.

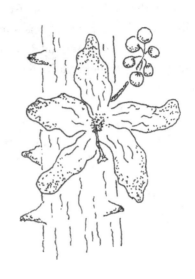

SILK FLOSS TREE—*Ceiba speciose*

The Silk Floss Tree

We can complain because rose bushes have thorns,
or rejoice because thorn bushes have roses.

—Abraham Lincoln

The silk floss tree is a bit like a rose bush on steroids. Its trunk thorns are so large and plentiful that the sight of them can be disturbing yet riveting. You want to avoid it and yet observe it closely. At least that was my reaction the first time I saw this tree.

There's one such tree in the Moorings garden. She stands alone on a small side path, with an open and clear area around it to allow her room to grow.

Until recently, the tough, thorny trunk made this tree seem masculine to me. But after visiting this tree several times, I see it as feminine. I've also heard of a Bolivian folktale about a goddess whose spirit is trapped in the silk floss tree. According to the legend, she was kidnapped by evil spirits who feared that the son she would bring into the world would punish them for evil deeds. The son was born and managed to escape, but the goddess remains trapped. That story explains the shape of the trunk with its mid-level fullness, bulge, or baby bump, as some would say. Perhaps being kidnapped, trapped, and left alone explains those angry looking thorns too.

Putting that little-known story aside, I've heard people call it the Beauty and the Beast tree as well as the medieval torture tree. *Ceiba speciose* is the botanical name for this specimen. I understand that in the eyes of some dendrophiles (tree lovers), the silk floss tree is the most beautiful tree in the world.

A close-up study of the trunk shows a verdant green bark beneath all the large, pointy thorns, which some observers see as spikes of medieval weaponry. But I just conversed with a woman who thought the thorns resemble seashells, not in shape and form, but in the tan and brown patterns on the surface of the thorns. She is among the fans of this tree.

I imagine a tongue-in-cheek sign here:

Caution to Dendrophiles.
Hugging a Silk Floss Tree Is Inadvisable.
Hug at Your Own Risk.

I stand with the mid-afternoon sun behind me, warming my back and casting long, ominous-looking shadows on the stucco wall behind this tree. The shadows suggest lifelessness or grief. And yet, as I shift my focus from the shadow to the trunk, I see verdancy and feel joy. Green is a color of healing, growing, hope, and new life. This tree is not only alive and healthy, it is growing, rapidly. Right now it is about ten to twelve feet tall. It can grow as tall as sixty feet.

Some silk floss trees don't flower until their twentieth year; the average age for flowering is eight. When the flowers do appear, they are bell or trumpet shaped, about the size of a large opened hand. The narrow, delicate petals in various shades of pink resemble those on some hibiscus plants. The lush, dense flowering looks like a celebration of life. Besides the beauty of its blooms, this tree has been valued for the silken floss in its pods. There was a time when this silk floss was used to fill life jackets.

I'm fond of this tree. I look forward to watching her mature as I continue to grow and change too. I wonder what lessons I'll see in her. This tree is already urging me to overlook superficial appearances, especially of thorniness. What seems hard in a person's exterior is quite often a protective shell for soft vulnerabilities. For some people, acting tough was required in childhood as a demonstration of strength. Then acting tough became a habitual way of being.

This reminds me of an older relative who once made a gruff first impression on me. Actually, he was a bit thorny all the time. Tall and thin, he had received the nickname "Spike" in his youth. This seemed to fit Uncle Spike's personality, too.

In many situations over the years, I had noticed a distinct difference in the way Uncle Spike and another uncle, Ray, treated children. Uncle Ray paid attention to what children said. He really listened. You could tell by the follow-up questions he asked and by how he applauded their accomplishments. He also genuinely enjoyed playing board games and engaging in contests with them. Uncle Spike did not relate as well to children. In fact, sometimes he was dismissive and gruff toward them. The personality he showed most often seemed covered in spikes and thorns to me. For a time, I judged him in "unclehood" as not good enough. But the problem with judging anyone as not good enough in any particular area is that it easily turns into judging the whole person as not good enough, period.

One summer day we had a serious plumbing problem in our home. The primitive basement in our century-old Victorian farmhouse was filling with sewage backup. As soon as he heard about it, Uncle Spike showed up in hip boots, prepared to help drain and clean the basement. He showed no reluctance or reservation about dealing with the mess or the stench. As I watched Uncle Spike descend the rickety wooden staircase, out of the blue Uncle Ray came to mind. While Uncle Ray was great with kids, he could not fix a thing and he would not have been willing to enter that basement to try.

Aha! I realized that I'd been comparing them and judging unfairly. Sometimes a person's strengths and talents may not be obvious. We all have different personalities and skill sets. And who am I to judge them in the first place?

Since that incident, I've had a clear mental picture of Uncle Spike in his hip boots after the job was done. He looks a little scruffy, but I see no thorns. In fact, he's got a sparkle of pride in being able to help. This image serves as a touchstone for me—a reminder to accept and appreciate people as they are. When I see someone's thorny, hard exterior, I can shift to look for the sparkle. Or now that I am getting to know the silk floss tree, I'll look for the flower and overlook the thorns.

The Beach Garden

The human soul is hungry for beauty;
we seek it everywhere... When we experience the
Beautiful, there is a sense of homecoming.

—John O'Donohue

It's early in the morning and I'm setting out for a beach walk to the Sanibel Lighthouse. A portable cassette player hangs from the belt on my shorts. After trudging through soft dunes, when I reach the hard-packed sand near the low-tide line, I turn left, toward the rising sun, and stretch. I set off. Keeping pace with the dance music playing, I step carefully around sandcastles and people bent at the waist, searching for shells in the classic "Sanibel stoop" posture.

With a row of dolphins arching just beyond the breaking waves and hundreds of birds nibbling at the waterline, it is easy to appreciate nature's beauty today.

About a half mile into my aerobic walk, I relax my pace to be in sync with the slower song playing.

I spot something above the high-tide line. Feeling drawn to it, I pull off my earphones and walk up from the water's edge to get a closer look. It's a simple, Zen-like garden made of leaves, twigs, and shells, some broken and some whole. In front of the garden, seaweed forms the word "Welcome."

I wonder if the greeting is intended for beach walkers, for imaginary creatures, or for the waves that could eventually claim this creation. The garden is not nearly as impressive as the sandcastles I have passed; it exudes beauty nonetheless. In fact, I like it so much that I decide to take a picture of it. That will have to wait until later though, because I am without my camera now. I replace my earphones and continue on my walk.

Later I return to the beach-debris garden with my camera and see a family of four setting up blankets quite near it. I introduce myself, learn that they are indeed the creators of the beach garden, and complement their creativity, especially their use of natural beach debris. The children thank me and tell me it was fun.

Their mother, stepping away from the rest of the family, gestures for me to follow. She says that she'd like to tell me the story of what happened.

She explains that she, her husband, and their two preteens had made a similar beach-art creation two days before. They'd had fun playing together, she said. When they stepped back and saw the result, they were pleased and hoped that beach walkers would enjoy it.

When they came out to the beach early the next morning, however, they discovered that it had been destroyed. Footprint evidence suggested that someone had stomped through their little creation, deliberately swishing and smashing his large feet around in it.

"When I saw the hurt on my children's faces, intense anger rose in me," says the mother. "Who would do such a thing? Why? I wanted to crush whoever did it. Fueled by fury, I marched down the beach, cursing the young man who I imagined did it, and, I have to admit, wishing him all sorts of harm." With a chuckle and a smile, she says, "I'm guessing that to passersby I looked quite fierce." She says that she had walked until her anger was spent, which brought her to the lighthouse at the end of the island. There she sat and thought, "What are my children learning from this? That some people are senselessly mean? To mistrust? To hate?"

"When I released the last of my anger, I felt something open," she says, "and the Dalai Lama came to mind. A sixth of the Tibetan population was killed in the 1950s. When he is asked about that holocaust, he responds with compassion for the people who did it. I then saw other possible lessons my children could learn from this experience."

She explains that when she had arrived back at the site, she had asked her family to forgive the perpetrator and to help her repair the garden. At first they resisted. So, she started to rebuild it herself. Soon, her husband joined her. Then, the daughter. Finally, the son. After a short time, they worked again in a playful way with ease and trust. They enjoyed another morning in the sunshine, arranging flotsam and jetsam in attractive patterns. They forgave the debris-garden vandal, wished him well, and agreed that what emerged this second time around was actually far prettier than what they had made the day before. She says, "When we finished, my son suggested adding the word 'Welcome' as a greeting and an acceptance of whatever would become of it."

For a moment I'm speechless, in awe of the mother's wisdom and compassion, and the son's maturity and understanding.

"Thank you," I manage to say. "Thank you for telling me your story. I admire you. What an example for your children. Not only to forgive him, but to wish him well. And for your son to suggest adding 'Welcome!' I'm so impressed. I will remember you."

I take several photos of the beach-debris garden before leaving the family alone to enjoy the rest of their day.

GIANT BAMBOO—*DENDROCALAMUS GIGANTEUS*

A Bamboo Concert

Notice that the stiffest tree is most easily
cracked, while the bamboo or willow
survives by bending with the wind.

—Bruce Lee

During today's garden stroll, I stop where the Wedding
Arch Path meets the Bamboo Corner. The paths here are
not actually named. They don't bear signage the way many
trees and plants do. But, out of necessity, my husband and
I have named a few. "Where are you?" he asks, calling my
cell phone from the beach. I look around, wondering how to
describe where I am.

"I'm on the Dockside Pool Path," I tell him. Another time he
tells me he's on the Avocado Path.

Where Wedding Arch Path meets Bamboo Corner is one
of my favorite locations here. I've stopped to give my full
attention to the bamboo concert that's happening in front
of me. This group performs regularly but at unscheduled
times, like impromptu jam sessions. What moves this group
to make music is the wind, of course. As the stalks sway,

they knock and clack; the leaves swish. I'm soothed by this natural percussion.

I sit on a nearby bench to listen. I watch. The bending bamboo reminds me of a fourth-grade girl in a school in Soldotna, Alaska. After reading *The Twelve Gifts of Birth* to the class, I asked the students to name one of the special abilities that make up their unique mix of talent. One hand shot up ahead of others, so I called on that child.

"I'm flexible," she said.

I've visited hundreds of classrooms and talked with thousands of students and I had never heard that response. My first thought was that she did yoga. Instead of simply, and wisely, asking her to tell me more, I went with my interpretation and asked if she did yoga.

"Yes," she said. "I do. But what I meant is that I'm good at accepting change and adapting to it. I'm willing to change my thinking and how I react to things."

Yes, a fourth grader said that! That insightful girl has become a touchstone for me. Like the bamboo in this garden, she is strong and flexible. She's able to bend without breaking in both her body and mind.

Nature's lessons are everywhere. I see the flexible aspect of strength in the bamboo and in the Alaska student.

The wind picks up and so does the music. The swishing, clacking, and knocking get louder. The bamboo bends lower. Along with keeping the concert going, the blowing wind keeps the no-see-um bugs away. I decide to stay and listen. And think. How accepting of change am I?

Time Travel

What nature delivers to us is never stale.
Because what nature creates has eternity in it.

—Isaac Bashevis Singer

Some days I feel drawn to a particular path. I call it the Time Traveler's Path, largely because, when I walk through this area, I feel as if I have one foot in the here and now and the other in the long ago. It takes no effort on my part to experience this; it just happens. I believe the traveler's palms are responsible. With their enormous, paddle-shaped leaves, they look old. They feel old. As a matter of fact, they're anciently old. Well, not these specific trees, but the species is. It's believed it has been around since 12,000 BCE.

Native to Madagascar, this exotic plant is actually not a palm. It's closer to banana trees and bird of paradise plants. Its proper name is *Ravenala madagascariensis*. How this huge, exotic plant acquired its common name has a few interesting claims behind it. The plant's giant leaf sheaths catch and hold a large amount of rainwater. Therefore, a traveler could count on obtaining water to drink from these plants. One legend says that if travelers or passersby make a wish while standing directly in front of the plant, the wish will come true. Another says that travelers can use the plant as a compass because the leaves tend to grow in an east–west direction.

I don't know how much truth those legends hold, but I do know that this intriguing area of the garden kindles imagining and dreaming. Almost every time I am here, I feel transported.

This section of the garden is miniscule compared to forests, yet I find here a measure of the silence and peace that tranquil forests offer. I suppose it's the ferns that blanket the ground beneath the palms that produce the cool, quiet forest effect. I can easily imagine a doe and her fawn resting here. I almost want to nap here myself. I wonder what dreams would come from sleeping on a bed of ferns beneath traveler's palms that reach thirty to fifty feet high.

I recognize that my habit of becoming imaginative in nature as an adult reaches back to my early childhood sense of wonder. I recall the untended yard on the side of the apartment building where I first lived. Instead of unsightly weeds, I saw wildflowers in a mini meadow. As young children often do, I perceived beauty in the tiny yellow and purple blooms on the tips of tall weeds.

Memories of the child I was become a time-travel vehicle. While one foot remains firmly grounded under the traveler's palms in the here and now, the other steps into the past. It's the early 1950s.

I'm sitting on the bottom brick step of my paternal grandparents' back stoop, watching birds drink and splash about in the birdbath in the center of their backyard. Shrubs of every size with flowers of every color edge the grassy square. Sometimes on days like this I would help my grandmother in her garden. As she nurtured her plants, she also nurtured my innate love of nature. But today isn't a gardening day. I breathe deeply and smile as I smell the strong scent of roses. I'm about five years old.

After the birds finish their baths, I open the screen door, enter a breezeway, then step up into the kitchen, where my grandmother stands at the kitchen table kneading dough. I smell stew simmering on the stove's back burner. I tell Grandma Gorda that I love the smell of her roses. She smiles and promises to send a bouquet home with me when my dad picks me up after work, which will be soon.

She covers the dough for its second rise and tells me it's her prayer time. We go into the living room and sit side by side on a loveseat as she retrieves her old, worn, black leather-bound prayer book.

I ask her to tell me, again, about her prayers. I ask this often.

"Of course, Charlanka." I liked the special way she sometimes said my name.

She tells me how she prays every day for me and each of our family members here in America. She names the twenty people who gather here every Sunday after Mass.

She shows me her prayer book, explaining that it's written in the Slovak language, the first language she spoke. I run my fingers over the strange-looking script. She teaches me a few Slovak words, we sing a Slovak song she's taught me before, and she reminds me how to count: *jeden, dva, tri…*

Then comes the best part. She prays for family back in the Old Country. Her family. Our family. My family. I've got relatives in the Old Country? Whenever I hear this, it leads me to feel "rich" in some way. I have cousins there. What are they like? I want to know names. Are there girls my age? Did they start kindergarten yet? She explains that all of our closest family is here in America. She doesn't know names and ages, but she knows that families grow, so I probably have cousins near my age there. And even if we never meet our distant relatives, we are still related to them—connected. Her caring about them and for them is a lesson for me, and a gift.

After praying for all of us living people, both here and there, she prays for all who died. Even though she says this prayer in English, many of the big words are unfamiliar to me. But I get an image of a special, holy light shining upon each of these people. I always get good feelings from how she prays for our dead relatives with love, especially her first two baby girls,

Anna and Helen. Sometimes she gets tears in her eyes when she prays for them. They are my aunts who are "like angels in heaven." And then, she prays for all of our family who are not even born yet. She prays for the children I will have some day. And their children. She even prays for the boy who will be my husband someday.

Here, under the traveler's palm, for a few minutes I bask in that memory. I feel as if she is with me.

With reluctance, I return to ordinary mindfulness. I'm in this garden in the here and now. At the same time, I'm feeling connected with my grandmother and her long-ago prayers.

Throughout their childhoods, when they were worried or upset, I sometimes reminded my children about those daily prayers, said long ago, like pebbles dropped into a pond of time. During any time of hurt or hardship, and every time I had any kind of accident or a near miss of one, it helped to remember them. I have a sense of them being truly present, having traveled through time to me, my husband, and my children, rippling over us. They ripple over my grandchildren now too—her great-great-grandchildren.

Grandma Gorda gave me the gift of seeing time and family differently. It was more like just a seed of seeing differently when she first presented it to me. But the seed "took" and became a seedling she watered and nurtured every time I saw her. It grew and bloomed. It continues to grow. I nurture it now. And I thank her. And I thank the traveler palms for the gift of this voyage through time.

Family Tree

Family: like branches on a tree, we all grow in different directions, yet our roots remain as one.

—Unknown

Long ago, when my grandmother taught me about praying for my relatives from the Old Country, I made a wish. I wished that I would someday get to know those relatives, some of them, at least. Maybe just one cousin my age. I was five years old. It was well before I ever saw a traveler palm tree or heard of its legend of making wishes come true. But now, right now, as I pray as my grandmother did, I realize that this wish has come true. It's something she would never have imagined. Nor would I have, really.

I haven't actually met them yet, not in real life, but I have communicated with several of them. I now know of and about some of my relatives who are distant in the world and at various distances on my family tree. Through a DNA testing and genealogy service, I've connected with a few distant cousins—1,327 so far! I learned that my ancestry includes many more ethnic groups than the Slovak, Lithuanian, and Polish I knew of growing up. I learned that I descend from people from many places in Europe, the Middle East, and Asia. I also learned that I have cousins living all over the world. Besides the United States, Slovakia, Lithuania, and Poland like I might have guessed, they live in Austria, Belarus, Canada, Croatia, Czechia, France, Germany, Greece, Hungary, Ireland, Italy, Mexico, Romania, Russia, Slovenia, Ukraine, the United Kingdom, and more... I never realized how widespread the branches of my family tree are.

As Grandma Gorda said, I may never meet them, but we are related. Now I care about and pray for them all.

Actually if you think about it, all families are connected. A family tree technically shows only direct ancestors—who someone is descended from, like parents and grandparents. But a genogram includes everyone who is related, including siblings, cousins, aunts and uncles, and so on. It shows how family trees connect. And they all eventually connect.

Yes, all of them.

All humans are related. So every single person I meet is my relative somehow. Every single person outside my immediate family is still a relative. Everyone else on this planet is a cousin of mine. And yours. Ours. I've heard it said that we are "all connected" or "all one." Well, in this way, we are. One family.

Words of Nature, Nature of Words

Let the rain kiss you. Let the rain beat upon your head with silver liquid drops. Let the rain sing you a lullaby.

—Langston Hughes

Rain is falling, ever so softly.

I once saw rain lighter than this, so light that it sparkled. It was as if there were tiny fairies, or fireflies, or sprites in each droplet.

I later learned it can happen when several conditions are present and lined up just so. In order to see the sparkles, a person has to be in the right place at the right time as well. I'm not seeing that magical rain at the present. But, right now, I'm in a place and moment of time that feels pretty wondrous too.

I'm sitting out on a screened and covered lanai which is surrounded by a wall of tall, colorful crotons. I feel as if I'm lovingly nestled and sheltered from the rain in a private outdoor room. This feeling is reminding me of a word I just learned. It stands for the blissful tranquility that some people feel when it's raining or storming and they are enjoying and appreciating indoor womb-like coziness. That word is *chrysalism*. There's a specific word for today's light rain too. It's *sirimiri* or "drizzle" in Spanish.

I've recently been discovering what are sometimes called "rare" nature words. They're rare because most people didn't learn them growing up and they are not in common use. It's not as if we need these words for everyday interactions; I may never use them in ordinary conversation. But I feel enriched just hearing them.

Psithurism is the sound of wind through trees. A low muffled sound like distant thunder, but thought to be caused by minor seismic tremors, is called *brontide*. The ultra-thin layer of glittering ice that sometimes forms on leaves, grass, and twigs is called *ammil*—a word that comes to us from the Devon dialect.

Another word I find fascinating is *yugen*. A Japanese word, said to be untranslatable, I have read that it suggests "that which is beyond what can be said," "a profound, mysterious sense of the beauty of the universe," "the depth of this world, as experienced through a cultivated imagination," and "an awareness of nature, the universe, or creation that triggers emotional responses too deep and powerful for words." Wow. That's quite a word.

I've been enjoying my word research, especially words related to nature. And philosophy. But I still feel a bit frustrated because there is a word that I believe we very much need, and it does not exist. At least, I've been searching for several decades, and I haven't found it yet. We don't yet have a definitive word for the powerful resources within us.

I refer to the twelve innate qualities that I write about as "gifts." Strength, beauty, courage, compassion, hope, joy, talent, imagination, reverence, wisdom, love, and faith are also referred to as capabilities, aptitudes, powers, essences, virtues, traits, and strengths. Because each of these words stands for a variety of things, not one of them directs our thinking precisely to these powerful "things." Calling them strengths, which they are, is inadequate and even confusing because one of the strengths is strength. "Gifts" seems trite to some people because the word usually means material presents. And yet they are gifts. Just as they are resources, capabilities, aptitudes, powers, essences, traits, and virtues.

For centuries, philosophers, prophets, and poets have been encouraging humanity to use strength, to have courage, to love, to be compassionate, to find and develop our talent, to hold hope, to recognize beauty...to mine what's in us for our own good and the good of all. These ideas have been discussed and debated through the ages. But what do we call this "stuff" in us that is so valuable and potentially very powerful? One thing I know it is, is a form of inner wealth.

Unlike outer, worldly wealth, we all have it. And cultivating it can enrich every area of our lives.

I feel strongly that this wealth, these innate qualities, these gifts, deserve to be better recognized, appreciated, and nurtured in ourselves and in each other. When we do this, we will be greatly enriched. And I look forward to someday finally finding the right word for them.

The gentle rain seems to be subsiding. Perhaps soon I'll be able to delight in *petrichor*, the earthy smell after rainfall, and enjoy *komorebi*, Japanese for sunlight streaming through trees.

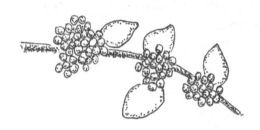

BEAUTYBERRY—*CALLICARPA AMERICANA*

Fruit of the Bloom

Discovering a hidden talent is actually a
very big deal... It can save the life of the real
you, the part of you that's special.

—Mira Kirshenbaum

After starting a load of laundry, I walk a short distance to visit
the beautyberry plant. Neither its little white flowers nor its
purplish berries are present at this time of the year, but I
like to visit anyway. I enjoy what I perceive as the easygoing
style, the generosity, the talent, and the caring nature of
this plant. She gently taught me a great deal just by being
her authentic self. I didn't always have this appreciation
for her. In fact, I thought she was boring and had little to
offer. But, boy, was I wrong. She offers plenty to humans
and animals. Since I have come to understand some things
about beautyberry, she has become a touchstone for me.

When it's not in bloom, at a glance, *Callicarpa americana*
can look like a ho-hum, nondescript bush. That was my
limited first impression of beautyberry. On my next visit
to the island, the berries were at their peak of ripeness and
beauty. As I walked by, I couldn't help but notice clusters of

stunning, shining, glistening magenta fruits. Several times during that week, I stopped and actually stared at them. I approached to get a better look and took pictures from various perspectives. In one tight, close-up view they looked like a bunch of miniature grapes. In another view they looked like gems worthy of use in jewelry making. From yet another perspective, I saw a portion of a long stem growing horizontally, appearing as a stiff rope clothesline, dangling berry clusters seemingly freshly washed and waiting to dry. I named that one Fruit of the Bloom.

Since then I've learned many more things about the beautyberry. This is now one of my favorite plants in this garden. Bear in mind, I say that about everything here. I don't really have a fixed favorite. At different times of day, in different seasons, when I'm engaged in some way with a plant, it's my favorite. I think each tree, plant, flower, vine, and shrub I've encountered here has been my favorite at some time, especially since I've been getting to know them all better.

Even in full flower, the tiny clusters of white blossoms could easily be missed by the human eye because they are so small and often hidden behind leaves. But when it's in bloom is another great time to photograph this beautyberry. Too bad its abstract characteristics and talents can't be captured on film. The frame would be filled with cheerfulness, generosity, helpfulness, easygoingness, and a host of talents.

In addition to appreciating this plant for the charm of its berries and flowers, I was impressed to learn that every part of the plant is useful to humans and has been used by numerous Native American tribes. The berries are made into jellies and jams, and the roots and leaves into tea. Individual parts and combinations of the bush have been used to treat a wide variety of ailments, including dysentery, stomach aches, fever, malaria, rheumatism, and colic. The leaves are said to deter mosquitoes and the bark can relieve itchy skin. It's also been used in dried floral arrangements and to make dyes.

I further learned that besides offering so many healthful and helpful uses to humans, it is a good source of food for many birds, including robins, cardinals, finches, mockingbirds, thrashers, and towhees. Armadillos, raccoons, opossums, squirrels, and gray foxes feast on the beautyberry too. Bees are very fond of the flowers which they find easily. Overall, the beautyberry is highly valuable to others. I had no idea that the plain, humble image she first presented to me was hiding so many talents!

Now I go out of my way to say hello to beautyberry. I appreciate her subtle beauty, but even more, I thank her for the lessons she taught me. Not to jump to conclusions based on a first impression. Not to judge a book by its cover. To remember that everyone has hidden talents, usually many. To remember that the expression of talents and beauty come at their own pace, in the right season, at their own time. To respect each unique form of life, its place in the world, and its value as a living being.

The Great Pretender

Be faithful in small things because it is
in them that your strength lies.

—Mother Teresa

On my way to the car for a quick trip to the grocery store, I pass a dwarf olive tree. I recall that, like the apple blossom cassia, despite its name, it bears no fruit. I wonder how these trees received their somewhat misleading names. It seems that they are pretenders, laying claim to traits they do not possess.

As I drive, I ponder how we all sometimes pretend…to be something we are not or to know something we don't. Suddenly I hear the song "The Great Pretender" in my mind. I was five years old when that song was released. I smile as I remember how my dad liked it and used to sing it while driving. Maybe often, but maybe only a few times. I don't remember clearly; things can impress upon us so strongly when we are children that we may give them more weight than they warrant. Musing, I wonder if he just liked the song or if he felt in some way like a great pretender himself.

I think about times in my life that I have been less than real, less than honest, or just not quite authentic.

Have you ever acted as if you were familiar with something, or someone, or a topic being discussed, when you really weren't? It may have been about a popular film, a historical incident, a recent news item, a celebrity, or the latest fashion. It might have been seriously important or completely trivial. Perhaps everyone in the conversation chimed in with knowledge about the topic, so what did you do? Act as if you knew about it in an effort to fit in? Remain silent and hope that no one would ask your opinion? Or did you courageously and unapologetically admit your unknowing and ask for information?

I think we've all pretended at times. I know that when I have, I have felt yucky, like an imposter, like I wasn't true to who I really am. It felt as if I betrayed myself. Perhaps only a subtle betrayal, but a betrayal, nonetheless. Trying to save face, to fit in with peers, to somehow measure up to others…it can lead us to let ourselves down. And sometimes it backfires and makes us look much worse.

As I glance at a restaurant I drive by, a funny story comes to mind. Years ago, when I was an executive recruiter (a.k.a. a headhunter), a coworker and I were looking for an exec for a food company. One afternoon she came over to my desk and asked, "Who is Bob Evans? Do you know a Bob Evans?"

"Bob Evans? You mean the restaurant chain?" I replied. "Oh no," she said, a look of alarm in her eyes. "Tell me it is named for Bob Evans and he is still alive," she pleaded, hanging her head. "I just told my client that I know Bob and I just talked with him the other day. No wonder he got silent and said he had to hurry to get to a meeting! I burned that bridge I guess..."

There and then we committed to no faking and no pretending. No flimflam to impress clients. No more stretching the truth to get past the doorkeepers. It was a turning point for both of us and became a touchstone for honesty, even in little things. If we don't keep our word about little things, why should we be trusted about big consequential things, with more at stake?

With age, I've become more and more comfortable admitting, "I don't know." I may feel some embarrassment, but it's slight now, and so much easier to bear than the feeling of pretending. It's actually empowering and nurtures peace and well-being. Being honest always does that. Being true. Authentic. We can even find humor in not being right.

Of course it can be difficult. That's human. We've been conditioned to appear successful, to be competitive, to be defensive of our image to others. But too often defensiveness leads to an argument, the deterioration of trust, or the degrading of relationships. Too often we become self-righteous, believing we must be right, so sure that things happened a certain way when, say, our spouse is equally sure it happened another way.

But it is truly so refreshing to come clean, to speak the truth, kindly, with genuine compassion and respect for ourselves and anyone else involved. I know that I respect people who come clean with me. This includes people I know well and anyone I have just met. I get to see the real person. I love them more for trusting me and feeling safe to be real with me. Don't we all want that? Anything else is just too much effort. Imagine if we all had the courage to be truthful about every little thing. Imagine if we each made others feel safe enough with us to be truly themselves. What a gift!

With a little practice, being fully honest gets easier. It gets easier to say, "I don't know," "oops," "I made a mistake," "I goofed," "I misspoke," "I was wrong," etc.

Entering the grocery store, I smile, hearing myself humming "The Great Pretender."

Wildflower Wisdom

I don't like formal gardens. I like wild nature.
It's just the wilderness instinct in me, I guess.

—Walt Disney

I like formal gardens. I see beauty in the patterns of plantings and the near perfection of the plants in such places. My eyes feast on their artistry. I appreciate and admire the intention and effort put into the planning and careful grooming of formal gardens.

Yes, I do like formal gardens, just as I occasionally enjoy a formal event, wearing a gown and jewelry, with my handsome husband in a tuxedo. But my true nature is more earthy, informal, and relaxed. More along the lines of sundresses, capris shorts, or even pajamas.

I much prefer wild nature to strictly controlled vegetation. It seems to me that in "natural" nature and in gardens that allow their inhabitants to be themselves, free, and at least a little bit wild, a palpable joy is present. I'm drawn to all gardens— botanical, English, wildflower, butterfly, vegetable, herb, windowsill, hydroponic, rock—but somehow I feel freest in a wild garden area. Relaxed. At ease.

I've just returned to the botanical garden at the Moorings after touring the grounds of Sanibel City Hall, where grooming is minimal. I took the Wildflower Walk, a one-and-a-half-hour guided tour of the areas all around City Hall. I came away from my wildflower tour feeling relaxed yet energized.

I had started off holding a stack of helpful handouts, a notebook, a pen, a water bottle, and my phone, keeping it handy to take photos of the dozens of beautiful plants. The guide was very knowledgeable and there was so much information I wanted to record. But after about ten minutes, I realized I had to make a choice. Do I continue to juggle all of these objects, try to write while standing, and ask the guide to repeat things I missed? Or do I forgo all that and simply enjoy the tour?

I decided to focus on the garden and let note-taking go. So I don't have any specific tips, anecdotes, or wildflower details. But it was worth it. Allowing myself to be in the moment and give most of my attention to the plants themselves made an enormous difference in my experience of the tour. Yes, I listened, watched, observed, and appreciated the information I was learning. However, the real pleasure I received came from my interaction with the plants and appreciation of the garden as a whole. It was beautiful and fragrant, a true feast for my eyes and my nose. But, there was more than that—sense, a feeling, from the plants. They felt like a congenial community in which all the native plants could relax in kinship and be themselves.

The area looked well cared for and exuded great health and well-being, peace, and harmony. Litter and fallen branches are routinely cleared away, but there is no regular pruning or shaping of plants. There is no rush to replace wilting flowers with fresh ones. The plants appeared to be allowed to live through their natural life span, at least more so than in other gardens. I felt a sense of ease. I could almost see it in the garden. The plants were not controlled, made to fit into perfect straight lines or boxes. Clearly there was little or no interfering with the plants' natural growth. I found this to be soothing, nurturing, restful, and uplifting. I felt enveloped by a soft yet joyful energy I imagined emanating from the cheerful plants.

Most of all, somehow, I felt reassured. It's as if authentic nature was encouraging me to be true to the real nature of me. Aiming for growth, but not trying to meet others' expectations, not trying to fit into another's mold. Accepting that I have an innate worth and a natural beauty as is. With self-compassion, self-love, and courage to just be me.

Here and There

I don't see the desert as barren at all; I see it as
full and ripe. It doesn't need to be flattered with
rain. It certainly needs rain, but it does with
what it has, and creates amazing beauty.

—Joy Harjo

The late afternoon low-tide conditions are perfect. The
gentlest of waves. No seaweed. A long sandbar ribbon on
which to walk. Golden sunlight. But after hours of near bliss—
reading, napping, beach walking, and floating—I'm ready for
some time out of the sun. I can almost feel the cool shower
water and smell the coconut-scented shampoo. I look forward
to slipping into a clean, loose, cotton sundress and sipping a
cold drink. I leave the beach via the arch bridge.

At the other end, I'm about to step onto the brick walkway,
near the first foot-washing bench, when something in my
periphery on the left catches my attention. Cacti? I see a
cluster of succulents, a very large agave, and a much larger
prickly pear cactus. What? I don't recall ever seeing this little
desert landscaping area. Perhaps because it is not along the
garden paths, I somehow never noticed it. Although I'm

looking forward to my after-beach comforts, I feel as if I must see these plants. Even though I'll be meeting them for the first time, they already feel like "forever friends." They are relatives of plants I have loved, still love, and miss, since moving from Arizona to Florida. It seems like it would be thoughtless, insensitive, and rude not to visit them.

When Frank and I moved from New York to Arizona, I was prepared to like the environment there, to adjust to it, but I didn't think I'd love the desert and the Southwest. Until we moved there, I had the impression the desert was harsh, dry, and barren. I was certain I would miss the forests, trees, and green hills of the Northeast, and being near the sea.

But the desert's awesome beauty and power took me by surprise and won me over. I soon discovered that it's not barren at all. In fact, it's rich with vegetation well suited to its conditions. I'm still amazed with how well it does with the little water it gets. And when it gets a little more water, wow! The blooming desert becomes a giant flower garden. Its beauty at times brought me to tears.

Seeing these desert plants now on Sanibel, I'm feeling reconnected with the Southwest.

I step closer to the succulent plants, greet them with an "I'm so happy to see you" smile, and look closely at each one. I especially enjoy examining the prickly pear cactus, because it was one of my favorites in Arizona. It still delights me. Its pads, also called paddles, are like wide, flat, thick leaves. To me they look like green oven mitts.

Although this desert area seems foreign to me now while I'm on this tropical island, it's a welcome sight. I'm feeling joyfully surprised, like one time when, traveling through an airport, I saw a dear school friend, who had been a best friend, but who I hadn't seen since a long-ago reunion. With only a little time to reconnect, having to hurry to different gates, that friend and I held each other in a long tight hug. I almost want to hug this whole group of cacti. But I'll never do that again! Well, I didn't exactly embrace a cactus. One jumped up onto me and wouldn't let go.

Shortly after we had moved to Phoenix, we took our first desert hike. It was more of a walk on one of the simplest, easiest, shortest trails on South Mountain, a part of the City of Phoenix's extensive park system. Neither of us had hiking boots yet. We were newbies. But we wore thick-soled sports shoes and socks, shorts, and hats, and brought water bottles. So we thought we were quite well prepared. And we were, mostly.

We were moving along at a moderate pace when Frank suggested we cut across a small natural area to where the trail looped around and continued. I don't know why he was in a hurry. If we stayed on the trail, we would probably have arrived at that spot across from us within only five minutes.

I wanted to keep to the trail, but Frank saw no problem with cutting across. There were no signs indicating not to. There were no large bushes between here and there blocking the way, just small short scrubby plants we could easily get around, so I said, "Oh, all right. Go ahead. I'll follow you." Then, about halfway across I screamed. I hadn't noticed anything come up behind me, but it must have been a coyote, I thought, that had taken a large bite out of my calf. I yowled and began to turn, ready to...I don't know...scare him off with my water bottle? Frank ran back to me, deep concern on his face. "What happened? What's wrong?" he was asking as he neared and looked at me.

Couldn't he see the animal somewhere behind me?

"Something bit me!" I yelled.

When I turned to look, I was shocked. Where was the animal? How had it disappeared so quickly? My leg hurt terribly, as it would after a large deep bite.

After glancing around and seeing no animal, Frank looked at my leg and saw that what had "bitten" me was still attached. It was a round cactus a little smaller than a tennis ball. We learned later that it's called a jumping cholla. Due to its fragile stem and light weight, a segment of a cholla is easily detached from the main plant by the slightest brush of a passing animal or person. Its protruding quills attach to the unsuspecting passerby, thereby allowing the plant to hitch a ride to another location as part of its reproductive process. A fascinating plant, its official name is *Cylindropuntia fulgida*. At the time, I didn't care what kind of cactus it was. I just wanted it to stop torturing me.

"How can this hurt this much?!"

Even though I did not say, "I told you so," Frank felt bad that cutting across was not only his idea, but that he had almost insisted. So, with repeated apologies, he bent down to remove the cactus. It then proceeded to "bite" him as well. As he tried to pull the cactus off my leg, some of its quills stuck into his hand. He tried to remove those with his other hand. The more he tried, the more quills stuck to him, one hand and then the other. I think he even got one in his lip trying to remove them. By now we were all stuck together, the three of us. I thought surely we would have to make our way to the car with me walking upright and Frank bending over or walking on his knees. However we did it, it would not be easy.

Finally, somehow, he managed to free both of us from the bulk of what we by then thought of as a deranged, possibly rabid cactus. But we still had to limp our way back to the trailhead, with hundreds of barbed quills embedded in our skin. With each step it felt like they somehow burrowed in deeper.

Once home, with tweezers, patience, and time, we succeeded in removing all the quills. Despite that introduction to Arizona, we lived there for more than fifteen years, loving it.

Oh, that memory... What was not funny then is hysterical to me now. Standing next to the prickly pear cactus, I'm laughing out loud as I picture Frank and me and that jumping cholla. I'm almost giddy.

A family walks by, a large one. My laughter gets their attention. I turn away slightly and try to stop, but it's way too late for that. I'm almost doubled over with uncontrollable laughter. I see a variety of expressions on faces young and old. A few teenagers, some adults, and young kids. Seeing their puzzled expressions makes me laugh even louder. Two little kids start laughing with me. They don't know why I'm laughing, but they laugh anyway. At their age, they still laugh for the fun of it, and much more often and more easily than adults laugh. But their giggles are contagious. An older woman laughs too. I think a few more join in. I can't even look. I have to rush to the closest rest rooms. I hope the "Gulls" isn't occupied. If it is, I'm using the "Buoys."

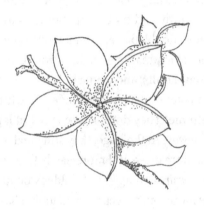

PLUMERIA—*PLUMERIA RUBRA*

Enduring Debby

We all have life storms, and when we get the rough
times and we recover from them, we should celebrate
that we got through it. No matter how bad it may seem,
there's always something beautiful that you can find.

—Mattie Stepanek

This isn't the Sanibel vacation my husband and I had
anticipated. Before coming, I pictured walking along the
beach, strolling on the sandbar at low tide, and meandering
through the garden. These are a few of our favorite things,
along with watching birds, waves, and the western sky as the
sun sinks into the sea. Reading, napping under an umbrella,
sharing a bucket of shrimp at our favorite restaurant, and just
being together round out our ideal beach vacation. Of course,
there's being in the warm gulf too—bobbing on gentle waves,
floating, treading, and talking. In all my imaginings—and his
too—the weather was perfect, as it had been for most of our
past trips here.

On the way down we decided we'd finally take a day cruise
to Useppa. We've heard that pods of dolphins often swim
and cavort alongside the boat and that a few hours on the

private, historic island feel like being transported to another place and time.

None of that is happening.

Although the forecast for the week ahead was less than ideal when we started our drive, and subtle signs of an offshore storm appeared about the same time we checked in, we saw no reason not to stay and make the most of this vacation time. Plus, there was no talk of severe weather nor possible evacuation. In fact, the storm was expected to turn seaward, away from the coast.

By the time we unpacked the car, it was already late afternoon. The sky was getting dark early for a June day, so we took just a short beach walk, during which my hat was snatched by a gust of wind. I watched it rise, fly, flip, and crash into wet sand. At least it wasn't into the waves; I retrieved it after a comical chase.

Storm Debby arrived the next day. Since then we've been inside for days, watching from behind glass doors blurred with salt, sand, water, and wet leaves. Occasionally, when the howling wind and heavy rain abate for a short time, we slide open the door and step onto the balcony to better see the turbulent gulf. The average wind speed is about forty miles per hour but gusts are as high as sixty-five m.p.h. Palms are bent to the ground.

We lose power a few times, and the condo darkens. We light candles and sit close, to each other and to the candles.

Witnessing this storm, enduring it together, is somewhat romantic. We know that many people are experiencing this storm, some in nearby condos. Nevertheless, while the storm rages, it seems like the two of us are very much alone.

This is not the vacation we envisioned, but it's riveting, galvanizing, and soul-stirring. We're feeling present and connected as we care for each other's comfort and safety. I believe we're safe. But there is a hint of danger, just enough to shake loose any sense of taking life or each other for granted.

We've been through plenty of various kinds of storms including metaphorical storms (life storms, marital storms), sandstorms, and snowstorms.

I'm noticing how natural and easy it is for us to turn toward each other, support each other, and become closer during weather-related storms. But life storms? We've held tight in some and not in others. I find it understandable yet sad. Life and relationship turbulences are often flooded with painful feelings and clouded with confusion. Heartache, fear, grief, regrets, or anger can act like driving winds that feed storms and blow things apart. It seems so wise to brave those winds with courage, compassion, and unity. In this tenderized moment, it almost seems easy. Ha. I know this is idealistic. But, whatever storms may be over the horizon for Frank and me, I intend to remember this feeling and to approach each storm with a spirit of cooperative adventure. I also hope storms are few and far between.

The power comes back. Lights go on. We blow out the candles and step onto the balcony. Palms are whipping and bending, but not so low. The rain has let up to almost a drizzle. We see people on the bridge which gently arches over the swale of natural vegetation paralleling the beach. This buffer area borders the sand and, in addition to being a beautiful conservation area, serves to protect areas farther inland from unusually high waves. Although the beach is underwater, waves are rushing no farther than to the edge of that densely packed area of sea grapes, beach daisies, and grasses. We decide to venture out and stand on the bridge with other spectators.

Ghastly green waves rise from roiling water that's charging in every direction. We watch for a while, then turn back and take a garden path before returning to the condo we are renting.

The garden is messy—branches, leaves, and flower petals are strewn everywhere—and yet I feel a pleasant peacefulness in it. Maybe it's because I'm away from the ferocious sea and there's a calm in the storm. But this garden calm seems independent of that.

I wonder if something we haven't yet discovered happens within the plants during a storm, a natural strengthening.

Maybe there's a plant form of faith that's part of their inherent will to survive. What if plants experience a camaraderie during stressful events—something remotely like the deepened connection my husband and I are feeling as a result of braving this storm?

Palms, bamboo, and young trees are still swaying but from the upright stature they've regained. Water is dripping, flowing, and running everywhere. Everything looks revitalized, energized.

We come to the bird bath near our building. Three perfect pink plumeria blossoms float like a work of art on the bowl's clear water. It seems like a sign that harmony and beauty are still present within the garden, despite the mess.

We're barely back in our condo when the high winds and heavy rains resume. Debby's mood fluctuates between rage and calm over the next few days. Sometime during the night, she departs.

The storm is finally over. The morning sky is blue. The sun is shining. The sea is calmer. Not calm enough for swimming yet; riptides are likely. We have one day left to do a few of the other things we anticipated before leaving tomorrow.

We begin with a beach walk. As usual, we turn left and head east. The sand is cleaner than one would expect. The receding waves from the last tide must have swept debris outward. One large, unidentifiable object lies ahead on our path. Waves wash over it, but it remains. Drawing near we're able to see what looks like a nature-made bridge. Curious, we pick up our pace.

It appears that waves have washed to shore two large logs of the same length and placed them parallel to each other about four feet apart. In the space between the logs, a cache of shells has collected, resembling a treasure trove for collectors. To me it seems like a bridge, maybe not over troubled water, but formed by troubled water.

I take a picture of this beautiful thing we've found formed by the storm. I'll keep it as a reminder that beauty can emerge from chaos and of my intention to stay side by side with my loved ones in all life's storms.

First Time Alone

The present moment is filled with joy and
happiness. If you are attentive, you will see it.

—Thich Nhat Hanh

It's the first day of my first solitary stay on Sanibel Island.
I'm here for a vacation but also to do some writing. I'm not
relaxed. Not yet. Like the average person, it usually takes
three or four days for me to release muscle and mind tensions
and shift into a vacation state of mind.

I decide to do something new this morning, at least different,
something I have not done for years—ride a bicycle. It's not a
daring adventure. Nevertheless, I feel an uplift at the prospect
of coasting along the smooth, flat bike paths. So, instead of
taking a beach walk or a dip in the pool, I head to the rack
where bikes are available for rent.

Striding purposefully in that direction, I'm almost oblivious
to my beautiful surroundings until a fluttering black butterfly
appears in front of me. When she circles my head in what
seems like an invitation to play, I laugh out loud. The garden
suddenly looks lighter. My vision sharpens, and my awareness

widens. I'm seeing the many shades of green, the bright pops of reds, and all the floral colors I always appreciate here. I'm smelling...mmm. Gardenias. My dad's favorite flower.

With the help of Black Butterfly, I've been transported to another state of mind. I'm in the present moment.

That butterfly is still with me. Or rather, I'm with her as she moves along. She leads and I follow. Ahead I see a cloud of wings like hers flitting around a huge bush. I think it's lantana; I don't know for sure. But I do know that I've never seen so many butterflies in one place. There must be twenty or thirty of them! Maybe more. I try to count them, but with all their movement, I can't keep the ones counted separate from the ones to be counted. So I give up counting and just watch.

Carefully, I step into their cloud. I stand still. I barely breathe. They don't disperse. They are so close that I am able to see luminous blue swatches on their black wings. I watch their fluttering, humbled, amazed that they stay so near for so long. I'm grateful. In awe.

The butterflies begin to disperse. A few remain. I linger for several moments. I'm reluctant to leave this spot which now feels magical.

My mood has shifted. My breathing has deepened. I'm relaxed. I choose to stay in the garden today, hoping to prolong this experience of wonder.

Black Swallowtail

Don't judge each day by the harvest you
reap but by the seeds that you plant.

—Robert Louis Stevenson

I learn that my fluttering friends are black swallowtails
and that the life span of that species is ten to twelve days.
Although we humans can expect to be here on earth many
more days than the black swallowtail—the average lifespan
for us is 27,375 days—our days are numbered too.

Where along its life span was the particular butterfly who
led me through a multitude of butterflies to another state of
mind? I wonder. Where along my life span am I? How well
am I living this once-in-all-eternity day? I recommit to living
each day more mindfully.

Thank you, Black Swallowtail.

From Holey to Holy

Things are not always what they seem...

—Phaedrus

Are things ever what they seem?

I learn that the holey leaves I see on milkweed and other plants are not diseased, as I first thought.

When I hear that a variety of caterpillars feed on those host plants and I understand that those leaves nurture the robust butterfly population here, those holey leaves become holy for me. Not ugly, but sacred and beautiful.

No-See-Ums

If you think you are too small to make a
difference, try sleeping with a mosquito.

—Dalai Lama

After a few wintery-for-Florida days, the temperature has
risen, almost reaching January's average high of seventy-five
degrees. It's a comfortable seventy-one, and I'm heading out
for an afternoon stroll. My intention is simply to notice.

First I notice the shampoo ginger plant and the miracle berry
bush just outside our door. I notice tiny white berries on a
bush whose name I don't know. I notice a giant staghorn fern
hanging from a royal poinciana. I notice a tree orchid and a
faint sweet fragrance. I appreciate the sights and smells as I
amble along.

I end up at Bamboo Corner, one of my favorite places. For
years I have enjoyed the knocking, swishing, and drumming
sounds made here. Two days ago I heard that, in addition to
the percussion song the stalks play together, each individual
stalk plays the sound of the sea. Experience proved it to be
true. The lulling wave music was loud and clear the other

day, almost mesmerizing, as if orchestrated in partnership with the wind.

I wrap my hand around a cool stalk and bring it toward my ear as I lean in and meet it halfway. I listen, straining to hear something. It's there. A sound. But it is very faint. Which makes sense because at the moment there's no wind.

No wind!? Oh no. Are they out? Biting? I don't know. If they are, I won't know until around six hours from now, or later, in the middle of the night. I don't see any. That's why they're called no-see-ums. I know enough to avoid them at their peak feeding times of dawn and dusk. But I had forgotten that they always seem to be out snacking on windless days, completely still days, like this one.

Maybe they are not as prolific during these cooler months. I'm usually here during the span of time from May through November, the nonpeak or off-season time. Oh, I think I felt a bite. Maybe it's just the thought of them and the power of suggestion. Just in case, I head back toward the condo.

I've been stung by one bee and two scorpions and bitten by countless mosquitoes over the years. A few months ago, I accidentally disturbed some red ants and in a matter of seconds received ten or so bites on my left foot. None of these are pleasant. But I think there is something especially insidious about no-see-ums.

First, you don't see them. Hence, the name. They are also called midges, punkies, hop-a-longs, and sand flies. These insects from the Ceratopogonidae family measure about a millimeter long. Varieties can be found throughout the world.

As I hurry back, I wonder: are no-see-ums among the gifts in this garden? What lessons might they offer?

Hmm...they are small things that take a bite out of us, unnoticed. Perhaps they can represent the little things that can take a bite out of our health and well-being, or our peace and prosperity. Those little things that we don't recognize at the time but feel the effects of later. I know to avoid dawn, dusk, and windless days in no-see-um territories in order to reduce the likelihood of being bitten. I realize that it is wise to become aware of and to avoid conditions that can diminish well-being. Or safety, comfort, health, wealth, or relationship harmony.

But perhaps the best lesson offered by no-see-ums is that tiny things can have enormous effects. Something as small as 0.03 inches long can impact a person's life for several weeks with itching. Some people experience pain from them. Sometimes they leave permanent marks.

As I enter my condo, I think about other small things that can have big impacts. I focus my thoughts on nice things though. Positive things. Several examples come to mind.

I remember reading about a six-year-old boy in Canada, Ryan Hreljac, who, upon learning that many people in Africa have difficulty accessing clean water, began raising money doing household chores so that he could help. He thought it would take seventy dollars. When he learned it was going to take more, he didn't give up. Within one year, through chores and fundraising, he had raised two thousand dollars, enough to build one well. He didn't stop there. He kept raising money. In two years, he was up to $61,000. This compassionate little child ended up starting a foundation which, over the years, has brought clean water to hundreds of thousands of people around the world.

And how much impact could one raspberry make? In her memoir, Gerda Weissmann Klein describes its significance. She was imprisoned in horrible conditions in a death camp during the Holocaust. Suffering was the norm: little food, bleak conditions, harsh work, and cruel treatment. One day a fellow prisoner, her childhood friend Ilse, chanced to find a raspberry in a ditch in the camp. Ilse carefully put it in her pocket and kept it hidden during the whole, long day of hard labor. At the end of the day she gave this rare tiny treasure to her best friend, Gerda, presented majestically on a leaf. I try to imagine the immense gratitude and joy the girls shared in this experience. This true story about one berry has touched millions of hearts around the world.

Size—our size, or the size of a gesture—doesn't matter when it comes to making a difference. As Mother Teresa said, we can all do small things with great love. Every day, many opportunities lie in wait. Some are so small that, like no-see-ums, they go unnoticed. Some, like a dusty, imperfect raspberry in a ditch, are spotted but may be passed over. Others we recognize. One example is letting a driver merge in front of us in heavy, bumper-to-bumper traffic. We can be that considerate driver who stops, smiles, and waves the other driver into the flow. Sometimes a gesture like this can change the course of someone's entire day, leading to them being kinder to others that day, resulting in ripples of kindness moving on to others from there. Many lives can be touched by one small action. No, smallness does not equate to powerlessness. Just think of the no-see-ums.

Down but Not Out

We don't even know how strong we are until we are forced to bring that hidden strength forward... The human capacity for survival and renewal is awesome.

—Isabel Allende

Today I'm exploring The Bailey Homestead Preserve and the demonstration gardens at the Native Garden Center here.

The first thing I see as I walk from the parking lot through the fence opening is a massive tree lying prone on the ground. Why hasn't it been cleared away? The downed tree confuses me, but also beckons. I'm curious. I approach and stare. As I walk slowly around the tree, I'm reminded of a piece of jarring modern sculpture that can be hung multiple ways. Front to back, back to front, upside down. There doesn't seem to be a clear top or bottom to this monstrosity. It's like no other plant I have seen before. As I walk around it again, stopping and looking more closely this time, I begin to recognize awesome things about this tree. For one thing, it is alive and growing. How did I not notice this earlier? I see roots going into the ground. Just a few from where the root base is closest to the dirt. When it fell, apparently a few roots held their grip. Branches protrude in sharp right angles from the trunk. Of course, I suppose that's natural now. They are reaching straight up toward the sun. A few moments ago, I thought these were all vines and shrubs growing near the tree, using this downed tree as a scaffold to clasp and climb. Now I see it's a strangler fig. I wonder how old it is. And when did it fall? There's a lot of other activity going on here—in, on, and around this tree. It's a home and place to play. Many geckos, ants, and beetles populate it, and two feisty squirrels take turns chasing each other over and around it.

I'm most in awe of the massive upturned root base. It's stunning—almost chilling to witness.

I think of how I might measure it, and Leonardo da Vinci's *Vitruvian Man* comes to mind. I can picture the drawing of the man with outstretched arms and legs in a circle, with four legs and arms instead of two, suggesting movement, maybe stretching. If I stretched out as far as I could, it wouldn't be enough to reach and measure the diameter of this tree. It's big. Really big. And the rest of it, the totality of the trunk and branches is so much bigger, sprawling. I wish my phone camera was able to capture the full size and expanse of it, but it simply can't.

How much water and nutrition could the few remaining anchored roots of this huge tree possibly draw up from the soil for its needs? I'm amazed it can still get enough. I sense incredible determination to survive, to grow, to continue to be. I recognize beauty now where before I saw deformity.

I'm told that winds from one hurricane pulled most of the tree's roots from the ground. A second hurricane brought the tree fully down. But, with those few roots still anchored, the tree survived.

I feel a kinship with this tree. It survived damage from two hurricanes; I survived two bouts of cancer and its treatments. Like this tree, my body's appearance has changed in many ways too. But I'm still me, determined to live well, with peace and well-being.

Truly, we all are in kinship with this tree. We've faced storms. We've weathered them. We have access to all the inherent strength we may ever need to survive and thrive. I know I'll remember this tree. Perhaps I'll bring it to mind the next time I need to draw on my strength. I'll picture those few roots, their tenacity, staying connected.

Dead but Not Done

Often when you think you're at the end of something,
you're at the beginning of something else.

—Fred Rogers

There are two of them. They look dangerous to me. Why aren't they taken down? "They" are two very tall coconut palm trees, dead ones.

Overall this garden is so well tended. Relaxed, but maintained. It's plain to see how loved it is. It seems to be in a near perfect balance of grooming and letting it be. It looks and feels natural, vibrantly alive, and even joyful.

But...those two very tall, dead, bare trees. The gardener in residence, who happens to be working nearby, sees me staring at one. She approaches me. With enthusiasm, she enlightens me.

"It's a snag," she says. "We've got two of them. You probably noticed the one across the street. I love them."

"Snags? Huh. I've never heard that term. I did see that other one. Aren't they dead palms?"

"Yup. But look at this one again. Closely."

"I...still see what looks like a dead tree."

"Look up to the very top," she says.

I shade my eyes from the sun and shift my gaze to where she's pointing.

Oh! A bird with a fish. "Is that an osprey?" I ask. "With his dinner?"

She smiles as she explains, "Yes, it is! These are dead trees, but extremely useful and valuable ones. This one is particularly helpful for our local osprey. They can dine here without being disturbed. You see, sometimes they are attacked by other birds while trying to eat elsewhere. So, this dead tree is a tall birdhouse, rest stop, and a restaurant."

She explains that snags are standing dead or dying trees that are both structural and functional components of habitats. They play many vital roles. They help birds and other animals raise and feed their young safely. Often there are dozens, maybe hundreds, of little holes up and down their trunks. Each hole serves as a home for an animal of some sort; a snag is like an apartment complex for them. In a natural forest, they also feed other life forms like mushrooms and lichen, and eventually decay and replenish the soil.

"While you're here, look often to the tops of these trees," she says. "You're likely to again see osprey taking their dinner there." She tells me a little about osprey becoming endangered. But that is turning around, thankfully, at least partly due to the efforts of people allowing dead trees like these to remain in place as they would in nature.

Once again, the garden gives me a lesson. This one seems to be: Don't assume, don't judge, don't jump to conclusions. Ask, listen, and learn.

Just a few moments ago these trees were unsightly and useless in my eyes. Now I recognize their value. And they truly appear different now, even beautiful. I am feeling grateful for these snags. And joyful and hopeful for the osprey too. I'll be sure to visit again tomorrow evening, right around osprey dinner time.

ROSEMARY—*Salvia rosmarinus*
BASIL—*Ocimum basilicum*
DILL—*Anethum graveolens*

Photo of the Day

Sense the blessings of the earth in the perfect arc
of a ripe tangerine, the taste of warm, fresh bread,
the circling flight of birds, the lavender color of
the sky shining in a late afternoon rain puddle...

—Jack Kornfield

The position of the sun tells me it's well past noon, and my
body tells me it's time to leave the blaze of the beach. I look
forward to walking through the cool comfort of the garden
on the way to our condo for a light midday meal.

After each long stretch of floating and bobbing in the waves of
the warm gulf water this morning, my husband and I rested
in the shade under our canopy. The beach is lined with them
today. In both directions, as far as the eye can see, tents,
umbrellas, and canopies like ours stand side by side like
festival vendors. I've never seen this beach this busy, but
then again, until now, we've never been here for the 4th of
July week.

I push myself up from the lounger, brush my swimsuit, shake
my towel, and trudge through the sand to the bridge that

arches over a natural growth area. I slip into my flip-flops and start across. At the apex I pause to take in the scene as Frank continues onward. In both directions, east and west, bridge after bridge provides safe crossing over the ribbon of wild vegetation that protects the coast from unusually high tides and storm surges. Nearby I spot a marsh bunny in a cluster of sea grapes. Two monarchs flutter around beach daisies. I imagine there are snakes in the thicket. None on this island are venomous. But still. I would not want to walk through this area where I also see prickly plants I have not yet identified.

Sweaty, sticky, salty, and sandy, I'm ready for a shower, so I move along. At the end of the bridge, muhly grass welcomes me onto the start of the tamed area of the grounds. Whether it's brown or pink in bloom, the light and feathery muhly grass tickles my joy. A little farther on, by the pool, I rinse off at the outdoor shower.

My thoughts turn to lunch. A perfectly ripe tomato, a ball of fresh mozzarella, and a bottle of extra virgin olive oil await us. But there's still the cool garden to enjoy, so I slow my pace when I step into the canopy of tall trees. A woman ahead of me hurries along the pathway. It appears that her awareness is not in the present moment.

I choose not to judge or assume anything about her. Often, I am not in the here and now myself, even after many "smell the roses" life lessons, including two bouts of cancer and three near-death experiences. There have been times of experiencing amazing grace and moments of feeling At-One. Usually such moments are short-lived but impactful, like my encounter with the black swallowtails here.

Synchronicities. I believe we've all had them, at least one. Mini-miracles. Maybe a major miracle. Lifeguiding lessons. Sometimes we miss them entirely. Luckily, there'll be more. We're all given opportunities to see beyond our ordinary vision. Every ordinary day we can see life with reverence and wonder.

I detour through the herb garden. There, with gratitude for simple pleasures, I pick sweet basil and oregano for our caprese salad.

When I arrive at our condo, I find Frank drizzling oil on the tomato and fresh mozzarella slices he's already cut and arranged. I wash and add the basil and oregano leaves and we carry our plates out to the lanai. The overhead slowly spinning fan moves just enough air to complement the gulf breeze. Fragrance from three plumeria blossoms floating in a small glass bowl delights me. Before we eat, I savor and honor this sensually sweet moment by imitating something my brother does. I snap a picture with my phone camera and declare the tabletop scene, my "photo of the day."

About ten years ago, my brother, Keith, started what he calls his "photo of the day practice." One day he suddenly felt, for no apparent reason, a deep grief for the months and years that had passed. So many moments. So many forgotten. He wondered how many were still ahead for him. It was not his birthday, nor had he experienced an accident or illness. His epiphany was not as much about facing his mortality as it was realizing that he has been taking his precious life for granted. He decided to change that. An idea came to him: every day he would take one photo in a purposeful way. He would pause, savor a moment, and honor it by capturing it. I was moved when he told me about this and took up the practice myself. We share photos at times. Some feature typical subjects—a stunning sunset, a flower in bloom, a pet at play. But many more depict the seemingly mundane moments of life. Keith's include a sunny egg frying in a pan, a just-poured glass of beer, and water flowing from a shower head. It's not about waiting for peak experiences; it's about appreciating ordinary moments.

Of course, there really are no ordinary moments. They're all magnificent. But we don't usually see life this way unless some significant change shakes us up. This "photo of the day" practice helps perpetuate the experience of reverence. Cultivating the habit of watchfulness for simple yet sacred moments can lead us to see many more of them.

Today a few flower blossoms, fresh-picked herbs, and a gentle breeze comprise this magnificent moment for me.

Smiley

> The next time you see a spiderweb, please, pause
> and look a little closer. You'll be seeing one of the
> most high-performance materials known to man.

—Cheryl Hayashi

In a shrub outside my door is an amazing silken web. Every time I pass by it, I glance over to see if Smiley is at home. Smiley is my nickname for the web's arthropod inhabitant. To me she looks cute, colorful, happy, and friendly. Almost cuddly.

This is not how I used to think of spiders.

Gasteracantha cancriformis, also known as the spiney-backed orb-weaver, has many attractive names including star spider and jewel box spider. Also known as a crab spider, her body does somewhat resemble a crab's. But there's always that bright, happy looking smile. So I don't think she ever appears crabby.

Her cheerful look is created by a series of circular marks in the shape of a smile on her ivory-colored, crabshaped back. Protruding outward from this crab shell-like "face"

are six coral-colored points. They remind me of the Statue of Liberty's seven-spiked crown. I think she's beautiful.

I know she's not the same Smiley year after year. This type of spider lives only about one year. I call her she because females are found any time of the year, while the males are usually seen only during October and November. Whether she is the same Smiley or a relative or descendant, I enjoy seeing her just the same.

The qualities I observe in the spiney-backed orb-weaver have led me to feel no fear of this spider. I must admit that some spiders still stir fear in me. But, overall, my perspective has changed drastically. Spiders are useful and fascinating. I no longer see them as worthless or something to hate, fear, or destroy. In my childhood I thought nothing of ending a spider's life.

But then came a point in my young adult life when I suddenly saw spiders differently. I recall the first of many times that our daughters screamed because there was a spider in the bathtub. "Daddy, Daddy! Get the spider!" they yelled. I don't know when my husband had his own wake-up experience about this, but he gently, calmly got that spider out of the bathtub and placed it outdoors. I had never before seen anyone do that.

I learned at that moment that there was no reason to hurt the spider; certainly there was no reason to kill the harmless arthropod. Nor did our young daughters need to fear the little guy. Little by little they let their fear of bugs go and began rescuing anything that did not belong in the house.

This acceptance of spiders and insects has been passed on to our grandchildren. In fact, upon starting school, whenever the occasion arose, they asked preschool teachers to please help them move insects outside instead of killing them.

Killing spiders and stepping on ants are among the things that can seem normal and unobjectionable to do unless we shift into a new understanding and awareness. It may seem like a small thing, but it gives me hope to learn how many other children and adults will now move insects out of harm's way or at least leave them alone.

I'm reminded of another behavior that used to feel normal to me, but not anymore. When I was a young child, in the 1950s, it was commonplace for people traveling in cars to toss paper wrappers, straws, used tissues, and all sorts of debris out of a moving car's window. For a time, I just assumed that disposing of garbage on the ground or in the great outdoors was what people do. And we did. Until we learned better. Many of us learned due to Lady Bird Johnson. In 1963 she made highway beautification her cause as First Lady. I recall the epiphany I felt when I first saw the "stop litter" ads and billboards. Of course! Why would it be acceptable to throw garbage on the ground? Now I find it shocking and incredibly disrespectful.

I wonder. My understanding of insects and of litter have changed so dramatically. What other ideas do I need to change? About what things have I not yet had a wake-up experience? How much more can I learn? I suspect it's a lot. I don't know what I don't know.

Joewood

The awareness of our own strength makes us modest.

–Paul Cezanne

Off the beaten path I find joewood. Also known as barbasco, cudjo wood, ironwood, and washwood, its proper name is *Jacquinia keyensis*. It is not along one of the garden's well-traveled walkways. I'm almost certain I've seen this shrub and smelled its flowers in other parts of the garden, but during this visit I've found only this one. It's in a cultivated section of plantings on a grassy mound near the beach in an area where many rabbits live. Although it is quite hardy, joewood is an endangered species. I believe that I've seen it in the garden around City Hall and maybe while kayaking through the mangrove backwaters of Ding Darling Wildlife Refuge. To help protect this plant by giving it more respect and attention, the city named joewood the official plant of Sanibel. I'm glad to have found it on this visit.

It is not showy or eye-catching. Its leaves are simple, with barrel- or paddle-like shapes. Its flowers are diminutive blossoms. In its simplicity I see poise, class, dignity, and humility. It is well worth taking the time to find in bloom.

For those attentive to their senses, joewood's scent—said to be heady and intoxicating like gardenia or jasmine—will likely draw them near for a closer look. They will then be rewarded with the sight of white, ivory, and creamy petals arranged in a star pattern in two rows with the smaller front row petals growing in between the larger petals in the back row. I get the sense that the joewood plant is comfortable with itself. It's at peace with its slow pace of growth and its small size, as well as its powerful beauty and scent. It neither apologizes nor boasts. It readily shares its visual beauty and sensational scent. To me, joewood seems to say, "Be comfortable with yourself. Be who you are, who you really are. Let yourself be. Don't compete. Don't judge. Just be."

ROYAL POINCIANA—*DELONIX REGIA*

Mistaken Identity

There are at least three reasons we should be slow to
judge: We sometimes don't have the full story.
We often project our own issues.
We usually regret it later.

—Michael Hyatt

I am exploring a stretch of the Shipley Trail for the first time.
Due to the summer heat and an empty water bottle, it has to
be a walk, not a hike, and a quick one at that. I erroneously
thought my thermos was full. I'm frustrated with myself for
not checking. Well, I better check more closely next time. I
quench my thirst with my last few ounces of water as I step
out of my air-conditioned car into a blast of heat. Whew.
It feels a lot like opening a hot oven. At least I have a hat
on my head.

Short, fast, and without water is not the way I typically
hike or walk. I like to take it all in and be attentive to what
I might experience as I look for beauty, birds, messages,
and surprises.

Looking to the right, I notice the old windmill on SCCF's Bailey Preserve. I admire it but keep walking.

What is that I see out of the corner of my eye? Gasping and jumping backward, I hear myself sputter, "snake!" Then…"SNAKE!"

Yikes! I almost stepped on it! Will it bite me? What kind is it? Is it venomous? No, there are no venomous snakes on Sanibel. Am I sure of that? What should I do? In an instant, my heart rate and breathing have increased, and my muscles have tensed.

Wait a minute… That's not a snake.

My body's fight or flight reaction was so swift and automatic that my eyes didn't have a chance to tell my brain to tell my body that they were mistaken. What I thought was a snake is a curved, almost two-foot-long seed pod from a royal poinciana tree. I am standing right under one. How could I have missed that? It is in bloom, too.

Sighing in relief, I think to myself, had I stopped or been more attentive to where I was walking instead of looking sideways at the windmill, I would not have leapt up, breathless with instant fear at what I barely glimpsed out of the corner of my eye.

I realize that I need to stop. Just stop. I stand still. I can't stop the adrenaline flow coursing through my body, but I can stop walking and stop letting my thoughts run wild. I can stop *not* paying attention. I can stop, focus my thoughts, and become present.

After a few centering breaths, I look up and see the amazing coral blooms and light feathery leaflets on the royal poinciana above me. A smile comes to my lips. My heart rate is starting to slow, but I know it will take at least thirty minutes for my body to recover from the conditions of this stress response. I laugh at myself, panicking in fear...of a seed pod.

Enjoying the shade of the tree, I think, *Wow, I was completely wrong in what I thought I was seeing. And my reaction was completely wrong for the circumstances.* I wonder how often my eyes deceive me. But is it our eyes or our minds that lead us to misinterpret what we see?

I think about this. Why did I assume it was a snake? Why did I instantly jump to that conclusion? Well, what I saw was curled; it looked serpentine. It was on the ground, which is where I have seen snakes before. And I am in a wildlife preserve. So...it could have been a snake. It fit my expectations.

Of course, I was mistaken, in part, because I wasn't paying attention. Not paying attention, not really listening or seeing objectively, allowing distractions to get in the way—all these things can lead to misunderstandings. I know this, but I forget sometimes. I get in a hurry or I am worried about something, so then I'm no longer in the present.

I have learned that our existing beliefs can cause mistakes too. We are all predisposed to think and feel certain ways. Most of the time we think what we are used to thinking and feel what we are used to feeling. Our minds want to fit what our eyes see and our ears hear into familiar patterns. Unfortunately, sometimes this means that we jump to a wrong conclusion and react inappropriately to the circumstances. Prejudices, assumptions, false and limiting beliefs...we don't usually even know we have them. But they affect our emotions, responses, choices, decisions, and actions, our whole lives, really.

Feeling my breathing slowing and my heart rate getting closer to normal, I realize that I feel drained after the adrenaline rush from fear. I decide to respect my body's needs. I need to get out of the heat and get some water. I turn back toward the car, deciding to return another day, better prepared and more focused.

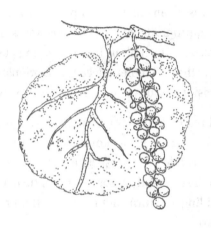

SEA GRAPE—*COCCOLOBA UVIFERA*

Sea Legs

You are imperfect, permanently, and
inevitably flawed. And you are beautiful.

—Amy Bloom

I'm headed over to the canal docks to meet Frank at the kayak
and canoe storage area. We plan to go for a short paddle
among the calm canals alternately lined with dense lush
mangroves and well-kept yards, docks, and boats. We don't
go often, but when we do, we usually spot something we
cannot see on the gulf side of the island. We always look
for manatees.

On my way, I stop for a quick visit with my pal Sea Legs.

Sea Legs is my nickname for a young sea grape plant growing
along the walkway that runs parallel to the canal. She is
another one of the many individual plant "friends" I've made
in the garden. Certain specific plants and trees seem to call
out to me when I pass by, something about their appearance
inviting me to imagine a background story and personality
for them. Yes, it may be a bit goofy, but I think, as with our

muscles, our imaginations can benefit from regular exercise. Especially as we age.

I met this friend on an earlier visit to the island. *Coccoloba uvifera*, commonly known as the sea grape tree, is a small evergreen tree or a shrub native to coastal beaches. It has white flowers, edible fruits that hang in clusters like grapes, and large, leathery, round or heart-shaped, red-veined leaves. I love the leaves; they are so sturdy that they used to be used as postcards. Sea grape trees are extremely hardy. They can withstand battering wind, pounding rain, and scorching sun, and they are drought resistant. I had seen these plants before, many of them. But this particular one jumped out at me. Her peculiar trunk gives her the appearance of standing on two scrawny, mottled legs. And she seems to be in a hurry to grow up, trying to look older by wearing high heels. She's not quite steady on them, so she leans a little as she struggles to keep balanced.

Sea Legs reminds me a little of a preteen or teen. I feel a kinship with this tree and a protectiveness toward her, as I would with a child or grandchild going through an awkward growth spurt.

The appearance of this tree with its skinny legs standing off-kilter touches my heart, but at the same time, this species is so strong and adaptable I'm inspired and impressed. I come to a kind of resolution about Sea Legs. She is going to be just fine. Better than fine. She has the makings of a great lady. Not a prima donna, but solid, resilient, and strong. She will hold her branches laden with leaves and grapes high with pride. I see her aging with grace, filling out well, and standing tall even if she leans a bit.

She clearly has imperfections, but she still has beauty. Ah. Not "but." *And* she has beauty. Her beauty is not despite the imperfection. Those so-called imperfections are part of what is beautiful. Seeing her in this way, I realize, is what the Japanese call *wabi-sabi*. Although difficult to translate, as I understand it, *wabi-sabi* is recognizing and appreciating beauty in transience, imperfection, and impermanence. It not only tolerates roughness, simplicity, and asymmetry, it celebrates them.

Seeing beauty in a *wabi-sabi* way feels right. It feels whole. Even holy.

Upside Down

O God, help me to believe the truth about
myself—no matter how beautiful it is!

—Macrina Wiederkehr

I read, write, and speak about the beauty and the other
gifts inherent in us. I believe it. My heart feels it as truth.
It resonates in my soul. But, honestly, despite all that,
somewhere in a corner of my subconsciousness there lingers
doubt. Somewhere in my mind there's a tiny hidden recording
whispering old false messages—stuck, stale ideas from my
past, from relatives, teachers, media, and coworkers—that
tell me I am not good enough, one way or another. Wrong.
Because I'm female, because I'm not an athlete, because my
hair is too long or too short, because I'm too short, because
I'm goofy, because I don't belong to this group or that group,
because my body doesn't look like a fashion model, because
I don't have a PhD or millions of dollars, because I'm not
as young as I used to be. Whatever the message was at the
time, it stuck.

I'm pretty sure most people have similar recordings, maybe
with a different beat or in a different key, but with the same

basic message, somehow saying that you or they or some people are not good enough. Ubiquitous societal pressures affect us all. They're subtle and not-so-subtle directives to fit into a mold of some sort, to present ourselves a particular way, to act or to look or to be a certain way.

I'm committed to knowing the real, beautiful truth about myself and all people so well that I can feel it in my bones. But I still need regular reminders that the old record playing softly but surely in the background is wrong, and to get myself to change the tune.

I experienced one such lesson in the garden at the Moorings on Sanibel. The teacher was a certain flower that continues to model the message each time I see her. I'm on my way for a visit now.

It's the start of a summertime one-week stay. This is my first garden walk since Frank and I arrived late yesterday. Without hesitation, I head directly to the purple passion flowers, *Passiflora*. I walk with a smile on my face and a skip in my step, excited to see my friend. Though I've forgotten what time of year it's in full flower, in my mind I envision the vine covered with blooms. Will it be? I believe so because I'm getting close and something in the air smells sweet, earthy, and a little salty. That's how I remember her scent. The purple passion flower is one of my favorites now. She taught me an important lesson and I'm grateful to her. I admire this flower...but it wasn't always this way. I didn't exactly get a great first impression.

The first time I saw the purple passion flower, I was taken aback.

I had discovered the vine that produces these unusual specimens during one of my open-ended meanders; they're the best kind in my opinion. Walking with no destination or time limits, I feel relaxed and comfortable, and generally have an open mind and heart. Or I think I do.

When I spotted the purple passion flower, an automatic reflex stopped me right there. I could go no further.

What the heck is that? What am I seeing? I thought to myself when I first saw it. It seemed upside down and inside out. I had to get a close look. I approached it slowly because I felt confused. Are my eyes deceiving me? Is my mind? What is that strange large thing on top? A monster-sized praying mantis? A lizard? Concluding it was not a large insect or reptile, I moved nearer.

I saw a purple, hair-like fringe that resembled what I've seen on Victorian lamps and 1920s flapper dresses. The flower looked completely open, as if its petals had been peeled like a banana, with the pieces hanging down to expose the fruit. Its sepals and petals appeared to be spread out and down, as if bending over backward to expose the working parts of the flower. It looked very weird to me. Not like a flower is supposed to look. Wrong.

Was I even feeling a little bit…afraid of this flower? Maybe. Some degree of fear often accompanies unknowns.

Right about then, I noticed a contradiction. Earlier on in the walk I was taking, I had felt open. Openminded and open-hearted. I thought I was open to life, to whatever I saw, wherever I went. I thought that was how I am, who I am—a strong open-to-life person. Especially after years of working on my "personal growth." But at that moment, my instant negative response to something as simple and inconsequential as an odd-looking flower was telling me point blank that maybe I wasn't as open as I thought.

Once I realized this, I almost scolded myself for my hubris, but I stopped myself in time, resisting the urge to self-judge. I took a breath. I relaxed. I turned off whatever record was playing in my mind telling me how a flower is "supposed" to look. And I looked again, closely. I examined its intricacies from top, bottom, and sides. With my heart, my eyes, and my mind really opened this time, I saw the flower with curiosity, fascination, and awe, as a beloved.

And she was magnificent.

The purple passion flower, bare to the world with her insides revealed and fringed hair flying outward, with her upside-down inside-out appearance, utterly refuses to comply with the standard image of a flower. And in so doing she unequivocally proclaims that she is good enough. That it's ok not to fit into any mold. It's ok to be different. Odd. Vulnerable. It's ok for her to show the world her up side, down side, inside, and outside. She is beautiful on all sides. And so are we.

Joie de Vivre

Joy is not in things, it is in us.

—Richard Wagner

I open the door, expecting to feel the familiar warmth and humidity that greets me here most mornings. When I step over the threshold, however, I'm embraced by an atmospheric condition I don't often experience. Calling it "balmy" or "perfect weather" doesn't adequately describe it. And it's more than the weather. There's something about the air, something in the air, that seems to contain healthfulness and a promise of good things to come. I think it's joy.

I've noticed that on days like this, people smile more than usual. Strangers say things like, "Isn't it a beautiful day?" and strike up conversations. Many little courtesies take place. It seems as if all is well in the world.

I remember a few early summer days like this when I was growing up in Linden, New Jersey. It would start with my mom going from window to window, lifting the sills, and leaving the windows wide open so outdoor air could come in and freshen every room. After a while the indoor air smelled as sweet as the cut grass and floral-scented outdoor air. I don't know how long that air-equalizing took because on those days I was out the door, around the corner, into Lawson Park, and on a swing probably before the last window was opened.

I was especially happy when I got the swing closest to the water fountain. On that swing, if you swung high enough, and if your legs were long enough, you could touch your toes to the tip of a low branch of one of the park's many shade trees. My toes never did touch that oak branch. But trying to touch a leaf amplified my joy, as did the "drop-of-the-stomach" sensation that followed.

I haven't been on a swing for many years. But I remember the feeling of being suspended for a moment in midair weightlessness at the peak of the arc, and then the drop, when the laws of gravity and motion pulled us—the swing and me—downward, backward, and up again. On such days, with so many positive conditions and experiences, everything about the world felt right, and the future looked joyfully bright.

With this childhood memory in mind, I take in a deep breath, exhale a sigh, and turn to the right on the path near my door. After a few paces, I stop and step aside for an elderly couple walking their small dog, Molly. I've forgotten their names but we exchange pleasantries while Molly sniffs my flip-flops and toes. When I laugh at the tickle, my head tilts back and my sight shifts upward. I spot an apple blossom cassia tree in full bloom and gasp audibly. It's as if I'm seeing that giant pink floral umbrella against the bright-blue sky for the first time. Its beauty magnifies the ambience of wellness in the air. I rave about the sight with Glenice and Bill—I've relearned their names—and snap a photo of the tree.

We rave about the weather too. Glenice reminisces about the scent of laundry drying outdoors on days like this. I share my New Jersey memory. Bill recollects similar days during his childhood in Pennsylvania. All this time Molly has been sitting patiently by Bill's side. But now she stands and looks up at Bill, saying with her big brown eyes she's ready to move on. What a sweet dog and amiable couple. We're about to part when we feel an increase in the mild movement of air that's been present. A moderate gust blows up from the gulf. We watch what happens when it reaches the apple blossom cassia tree. Profusions of pink petals rain down and carpet the ground.

"And that's why it's named a pink shower tree!" exclaims Glenice.

"I thought it was apple blossom cassia," I say.

"Oh, yes. It has that name too," says Glenice. "But it bears no apples."

"So how do you like them apples?" Bill chimes in and laughs at his joke.

"Come on, you old wise apple," says Glenice, threading her arm over the bend in Bill's elbow.

He places his hand on hers, with, "Ok, apple of my eye."

As they amble down the path, I hear him singing about little green apples in the summertime. They certainly have joie de vivre in their family. My encounter with them has nurtured my joy and enriched my soul. No wonder Molly looks like she's smiling all the time.

If I Had an Apple

If you have an apple and I have an apple and
we exchange these apples, then you and I will
still each have one apple. But if you have an
idea and I have an idea and we exchange these
ideas, then each of us will have two ideas.

—George Bernard Shaw

On my way to get a close-up look at the pink shower/apple blossom cassia tree from beneath the canopy, I chuckle about Bill's crisp apple quips. I admire the playful banter between Bill and Glenice and appreciate the joy the couple shared with me. My breezy conversation with them complements this "all's right with the world" balmy day.

I'm standing under the glorious dense canopy looking up. This view is a feast for the eyes too. The creamy ivory, bright pink, and deep rose-colored petals remind me of popcorn, bubble gum, and strawberries. The shape and form of each blossom looks like a kernel of corn popped open to extra-large fullness. There's no sign of apples. As Glenice said, no matter how many blossoms are on this tree or how many bees buzz in and out of them, this tree will never bear fruit.

I'm hoping for another gust of wind. I'd like to be standing here under the canopy when it showers a plenitude of petals. I'll wait. Chances are good it will happen again. There's been a regular series of puffs drifting up from the gulf.

While I wait, I touch the wide gray trunk. It's smooth and cool yet feels soft and solid. I think back to the couple I just left. Their long-term relationship seems soft yet solid. The warmth, grace, and ease of our exchange remains with me.

It gets me thinking about communication. I've noticed that when another person and I bring genuine interest, sincere respect, and a true desire to understand each other into conversations, the talk is fresh, clean, and clear. Like weather that invites the opening of all the windows in a house, it's as if the other person and I have opened windows to our hearts, minds, and souls.

It's not always easy to have conversations like that. In fact, it's usually not easy. In the exchange I just had with Glenice and Bill it was effortless, but that conversation was light and on a surface level. For deeper and more challenging conversations, most of us need to learn how to listen. What we commonly do isn't really listening. It's merely waiting and getting ready for our turn to speak. Sometimes we don't even wait. I know that I'm guilty of interrupting; I've been working to break that habit for a while. Even if we stay quiet and seem patient, we're often formulating a defense or an argument while the other person speaks. That may have its place...in courts and debate competitions.

But for success in relationships of all sorts, in everyday life, in real conversations, we must at least endeavor to understand others' viewpoints. And they need to hear and feel our understanding. Just as we need to feel theirs. Yes, it's a lot more work.

I used to judge or simply ignore when someone was expressing an idea that was contrary or offensive to how I see things. I'm sure I still do this sometimes, maybe often. But I have learned, at least in principle, that it is not useful to avoid topics that trigger discomfort, embarrassment, or frustration. Conflict isn't truly avoided that way, although it may be postponed. Now, when I remember to, if someone says something that clashes with my ideals and beliefs, I try to think of it as an opportunity. An opportunity to learn and even exchange ideas. I'll say something like, "Please help me understand your point of view," and listen (hopefully calmly!) to the person's response. It doesn't always work. But when it does, even a little, I feel an enormous sense of hope. I breathe more easily.

Ahh. Here it comes, I think. Yes. A gust. A gentle one. Enough to stir the blossoms. There's not a downpour, just a little shower. But it feels like a showering of blessings. I close my eyes and breathe deeply.

Bird-Watching, Bee-Watching

The bee collects honey from flowers in such a
way as to do the least damage or destruction to
them, and he leaves them whole, undamaged,
and fresh, just as he found them.

—St. Francis de Sales

Millions of people around the world watch birds. I'm one of
them. If you're reading this, you might be one too. I'm not
quite a birder. I'm just always ready to notice and observe
birds. Egrets, willets, plovers, and roseate spoonbills are
among the many birds to spot here. Osprey and bald eagles
too. Today I've been most fascinated with the prehistoric-
looking pelicans. But after a full morning of bird-watching
and beach walking, I've moved into the garden.

Now I'm watching bees. Standing in front of the *Cassia
javanica* tree, I'm far enough away to respect their space,
yet near enough to be awed. Beating at 230 times per second,
the bees' wings become a blur when they fly from one pink
blossom to the next. They enter the center of each flower with
what looks like purpose and care, taking what they need, and
giving something needed. In this moment the bees give me a

simple yet significant example of how to live in society and on the planet. Give some. Take some. Take only what you need.

APPLE BLOSSOM CASSIA LEAVES—
Cassia javanica

Seeing the Light

The leaves and the light are one.

—Albert Einstein

The bees' relationship with flowers is a harmonious balance. The flow, the giving and the partaking, seems, at this moment, sacramental to me. Holy. I see strength, wisdom, and beauty in these bees. I thank them for their graceful example and their gift of pollination. It's essential to the survival of many plants and to the animals and people who need those plants.

After feeling an almost sacred moment with them, I step back a pace.

As I do, my attention and awe shift from bees to leaves. The *Cassia javanica*'s leaves look a bit like ferns. I appreciate their design: twelve sets of side by side leaflets, all connecting to a center stem. As I look at the individual leaflets, the light changes; sunlight shines upon their backs. It's as if light is pouring into each leaf. They look brighter, warmer, greener, even happier. Is that possible? Happy leaves? They exude a sense of well-being. My smile widens. I think maybe my happiness is putting a happy filter over how I'm seeing now.

No matter. Due to the leaves' opaqueness, sunlight shines through each leaf. But it's so bright, it appears that light is coming from the leaves, present within them, not just shining through.

Their glow reminds me of lightning bugs. They always fascinated and thrilled me as a child. I remember waiting for the lightning bugs to come out on summer nights in New Jersey. Almost nightly, after I'd helped my mother do the dinner dishes and put them away, I'd sit on the front stoop. I'd look from the sky to the lawn, to the sky, to the lawn, anticipating both the first star to wish upon and the first lightning bug, not wanting to miss either's arrival. The lightning bugs, illuminated from within, were particularly wondrous. They felt magical. These shining leaves do too.

Namaste, Jai Bhagwan

People...are all walking around shining like the sun...
I suddenly saw the secret beauty of their hearts...
If only we could see each other that way all the time.

—Thomas Merton

I'm walking back to my condo after spending time at the beach and in the garden today. Light in leaves. Light in lightning bugs. Still feeling wonder and awe from my experiences with the bees and the leaves of the pink shower tree, I contemplate the idea of "light" within. What about light in people? There's a saying to "Let your light shine." What is that light within? What does it comprise? What does it represent? Gifts? Talent? Personality? Energy? Spirit? Consciousness? Love?

Many yoga, chi gong, and meditation sessions begin and end with a bow and the word *namaste*. Session leaders and teachers explain the meaning behind the gesture and the words. "The Light in me sees and honors the Light in you," or "the Divine in me greets the Divine in you."

Jai bhagwan has a similar meaning and intention. I remember first hearing this expression in the summer of 1990 when my college-age daughter and I went on a week-long yoga retreat program called The Art of Joyful Living. During orientation we learned we would be using this greeting before and after every session. It was said with hands in the prayer position and with a respectful small bow. At first most of us struggled a bit and laughed at our mispronunciations. Self-conscious chuckles continued for a session or two. But as the week progressed and together we learned, wrote, reflected, prayed, played, painted, shared stories, and listened, our hearts opened to ourselves and to one another. Deep respect was cultivated. Respect beyond respect, it contained love and reverence. And the slight bow and the "jai bhagwan" became sincere and profound. Even if we have experiences like that rarely, or even once in our lives, it can be life changing.

Dawn

I believe that imagination is stronger than knowledge...
That hope always triumphs over experience.
And I believe that love is stronger than death.

—Robert Fulghum

I'm standing in front of a bare plumeria tree alongside Marjorie from Chicago. We just met.

"Is it dead?" she asks.

"It looks dead, doesn't it?"

"Yes, especially surrounded by all this lushness," she says, sweeping an arm in an arch over a mass of vibrant greenery.

I tell Marjorie about the time I was standing right here a few years ago, thinking the same thing, when the gardener in residence at the time came by and said, "No. it's not dead," without my saying a word.

Marjorie laughs and asks if the gardener was a mind reader even though she caught the gist that apparently more than a few people have thought this tree, while dormant, is dead.

"I've seen other deciduous trees here out of season," she says. "With a few leaves hanging on, they look dormant but alive. This one really looks lifeless."

Marjorie goes on to tell me about a scratch test to see if there is green beneath a tree's bark, and that now and then a plant that looks really, really dead after a harsh winter comes to life again. "We often have to give things that look dead a chance," she says.

Marjorie gets my full attention, not only by what she says, but by the way she says it, as if she has more to add. Instead of responding, jumping in as I often do, I just wait. Marjorie draws in a breath and continues, "Sometimes, they are not dead. Sometimes, they are. At times, miracles seem to happen. And sometimes they really do."

I like Marjorie's company and would like to talk a little longer, especially since we started to go beneath the surface. I've never been one for small talk. I appreciate depth and authenticity in conversation, even with new acquaintances. I wonder if she wants to share something; I get that sense. I know that I would like to. Marjorie is wearing a wide-brimmed hat and comfortable sandals, as am I, and we both have water bottles with us. But we've been standing for quite some time, and in the sun, so I suggest a walk to the sanctuary garden.

"Marjorie, I have some photos of plumeria on my phone. I'd like to show you one. But, more than that, I'd like to share a story about a bird."

"I'd love to hear about the bird," she says. "And see the flower."

On the way to the sanctuary garden, I tell Marjorie that the flowers on the particular tree we just left are a vivid pink and that other plumeria flowers around the garden are white with yellow centers. I take a seat in the sanctuary garden, moving the chair a little closer to the one she's just chosen so that we can speak softly but still hear each other.

Marjorie waits patiently while I scroll through photos. Finally I find and show her a photo of the pink plumeria tree we thought looked dead. It's at the height of its full bloom. "Oh!" Marjorie exclaims at the striking abundance of flowers, "That is far from dead. It's so vibrant! It's almost quivering with life! It's gorgeous!"

Before I begin my story, I pause. The story I intend to share is one I've written about and told often, but not for some time. It is emotional for me, and I feel it deserves to be said from my heart, with respect and honor for its "characters," and with gratitude for my experience from it.

With a deep breath, I begin. "One day, many years ago, my husband and I noticed an unusual silence in the house upon returning home from a family party with our young daughters. Usually our zebra finch greeted us with song whenever the front door of our house opened. His name was Cubby. He had black, white, and gray stripes and a spot of bright orange—so colorful. And he was sweet and sociable. He never failed to imitate the squeaks of the door's hinges with squeaks of his own. That evening, however, we found him lying on the bottom of his cage, completely still.

"With sorrow, we comforted our crying daughters. We didn't want to diminish the importance of their or our grief, but we reminded them that he had lived well beyond the estimated lifespan of zebra finches and long after his mate had died. We said that we were thankful for having shared Cubby's life as long as we had. Because it was nighttime, we planned to bury him the next day.

"I went to his cage when I awoke at dawn. Although Cubby was still motionless on the bottom of the cage, he was upright, not lying on his side like the night before. With his head tucked down and his body pulled in toward its center, he took the form of downy, feathered ball. Although he seemed barely alive, clearly he was not dead. I was shocked!

"Now, if this happened today, I take the bird to the vet for help. But back then, we didn't know as much. We thought that vets were only for dogs, cats, and farm animals. So we let nature take its course.

"For days, Cubby stayed in that downy-ball position on the bottom of the cage. Then, slowly, he began to move about. We were amazed. We watched him recuperate. Truly in awe. After that for quite a time, Cubby lived quietly, sort of in a limited way, staying on the bottom of his cage. Of course he ate and drank and pooped like birds do. But he stayed only on the bottom of the cage. Finally, one day, he hopped up to a high perch. From that point on, he seemed like he was back to his old self, in all ways but one: he no longer sang.

"Months later, we came in the front door—I don't remember where we'd been—and we heard that familiar chirp! The girls rushed to see Cubby, and he continued to sing to them. For two more years, that special little zebra finch filled our home with song. I'm still so amazed when I think about it. That experience is a touchstone for me, a reminder. Of hope. And of inner hidden strength."

"Wow, that is remarkable. Thank you for telling me," Marjorie said.

"Well, there is more to the story," I replied.

"Oh, then please keep going," she said.

"A year after Cubby died, I got a terrible phone call from my brother in tears. My mother had had a severe stroke, and she had very little time left. So my husband, our daughters, and I flew to be with her. I remember crying the whole time on the airplane. When we got to the hospital, I hardly recognized my mom. I asked the doctor about her prognosis. He shook his head and lowered his eyes, avoiding mine. 'What are you telling me, doctor?' I asked. 'It was massive,' he said. 'You should prepare yourself.' I was devastated.

"Sometime during that restless night, as I cried and prayed for her peaceful passing, a voice within jolted me. *Remember Cubby*, it said. *Where there is life, there is hope*. I remembered how the innate life force of that little bird had directed his healing. So I stopped praying for my mom to pass peacefully and asked for her not to pass away at all. Well, at least, not yet. I began to pray for my mother to heal. I even tried to picture her as healthy, although she looked awful. After a while, I also prayed for peace when life would eventually leave her body, but I asked that it be far into the future.

"Well, despite her appearance and the dismal predictions from the intensive care staff, my mother, just like Cubby, pulled through! It took a while, but her healing kept progressing. In less than two years, my mother fully recovered. You wouldn't have known that she had had a stroke. I think she was healthier and certainly more active than before the stroke. Every day she walked three miles through a park near her home; that was completely new for her.

"She lived to see the graduations of her granddaughters from both high school and college. She lived to see the birth of a grandson. She even came here to Sanibel; I spent a week with her here. It was a wonderful week. Sometimes, when I visit, I walk by the condo where we stayed and pause at the door to say a little prayer and remember her. I can so easily picture our trip in my memory. We're standing together at the door, with me getting the key out of my beach bag. Or we're making lunch together or sitting on the lanai. When I stop at that door, I can almost feel her presence. I can almost smell her and feel her so near, right next to me."

I have tears in my eyes. Marjorie reaches over, touches my hand, and thanks me for sharing. I notice tears in her eyes, too, and I thank her for listening.

For a few moments we remain, sitting silently in the sanctuary garden, accompanied by the gentle sound of the fountain.

A burst of laughter from an approaching garden tour interrupts our tranquility.

We both stand.

"I haven't gone by that door yet on this trip. I think I'll walk there now," I say.

"Do you prefer to be alone?"

"No, I'd appreciate your company."

As we walk away, Marjorie begins, "I have a story to share too…"

Four-Leaf Clover

It is the little things which reveal the chapters of
the heart. It is the little attentions, the numerous
small incidents and simple courtesies of life,
that make up the sum of life's happiness.

—Ellen G. White

I come to a lush lawn area between the beach and the
buildings and notice a little patch of something softer and
darker green than the majority of the grass. I approach
and recognize that it is clover. Clover! I'm happy to see it;
I didn't realize it, but I miss it, like an old familiar friend. I
have not seen clover in a long time. I used to see it in every
lawn when I was growing up in New Jersey. I loved seeing
its white puffy flowers dot the yards and occasionally feed a
hungry bunny. Sadly, sometime over the years since then, the
clover plant was rebranded and labelled a "weed." It is quite
beneficial, though. For example, it is a nitrogen fixing plant,
so it feeds other plants around it. It also feeds wildlife. And,
in my opinion, it's simply more interesting than plain grass.
I remember spending many childhood hours in patches of

soft clover searching for the fabled four-leafed specimen for good luck. I never found one.

I was given one though, in 1999. One real four-leaf clover is taped into a guest book I used at book signings and readings back then. I brought that guest book with me everywhere I went during the year Frank and I traveled around the country in a motor home to market the first book I published. When I did readings and signed books, people wrote little notes or just signed their names. The four-leaf clover is unique in the book, which contains very few mementos other than writing, signatures, and a few photos.

An eight-year-old boy named Michael gave the clover to me when I visited a residential school for children with special emotional needs. After reading my book to the children, we discussed their twelve gifts—at what times they used strength, how they liked to use imagination, what special talents they saw in each other, and what compassion meant to them, for example. After that, I gave each child a small gift. It was a polished stone, to serve as a reminder of their inner strength, beauty, and all of the gifts, at any time they might forget that they have them inside themselves.

When we finished, this boy approached me and told me he wanted to give me something as a thank you. He had nothing in his hands so I couldn't imagine what it might be. He then reached into a pocket in his jeans, carefully fished around, and found what he was feeling for. Slowly he pulled his hand from his pocket and proudly handed me a wilted but perfect four-leaf clover.

Shocked and touched, I thanked him and told him that I would love to keep the memory of the four-leaf clover and that I would surely always remember both him and his generosity, but that I wanted him to keep the clover because they are so hard to find. I told him how I looked and looked for years, many, many times and never found one, so I wanted him to have the special one he found. But he insisted that I keep it and said he really wanted me to have this one for good luck.

I still have it. That boy must be almost thirty years old now. I have thought of him often over the years, hoping that he is having a good life—happy, healthy...and lucky too. And after two decades here I am again thinking about him and his small gift. Each time I have remembered it, it is as if I am given the gift anew. Not the gift of the clover itself, but the gift of a reminder, an example, of unconditional generosity. Of the innocent, joyful, loving attitude of a child. A child who was himself in difficult circumstances at that time.

We touched each other back then. Briefly, but meaningfully. Real impact and lasting effects in the lives of others can come from the smallest, simplest acts. We should not overlook or dismiss seemingly little gestures. Through them we touch lives in ways that can last a lifetime.

Turk's Cap Mallow

The world is full of magic things, patiently
waiting for our senses to grow sharper.

—W.B. Yeats

Occasionally when I meet someone for the first time, it feels as if I have known them before. I immediately feel comfortable with them. I don't know why; I perceive something indefinable but clear that attracts me to them. We just click. I experience that with plants and trees too.

For example, I recall a tree I passed each day walking to and from elementary school. It was a type of pine. From the first time I saw it, it felt magical—no, sacred—to me. It was extremely tall and the only one of its kind anywhere within sight. I called it the grandfather tree; it seemed wise and holy to me. I always stopped to look at it. I felt as if it somehow knew me and protected me.

Well, on the other side of this, there are friends I have made more slowly...people and plants. Sometimes it takes a while. Turk's cap mallow is one I've come to appreciate, really appreciate greatly, but only recently. At first it felt somehow disappointing. But now, after getting to know it, it's among my favorites.

Turk's cap mallow is the common name used for two different shrubs related to hibiscus. Their proper names are *Malvaviscus penduliflorus* and *Malvaviscus arboreus*. These shrubs go by many common names. Some are quite fanciful. Several alternate names for turk's cap mallow are: Drummond's turk's cap, turk's turban, sleepy mallow, sleeping hibiscus, wax mallow, Mexican apple, bleeding hearts, ladies teardrop, Scotchman's purse, and cardinal's hat. I find it in several places in the garden here. Today I'm visiting the one near the wedding arch.

The most distinctive feature of this perennial is that the flowers continuously look like they are just about to open, but they don't. When I first learned this, I felt sadness and even slight frustration. I almost wanted to pry them open, to help them along.

The flowers' perpetual closure seemed to mean that the plant would never reach its full potential. That its growth had been arrested at a fixed point in its development, never to fully bloom. Why would these flowers do this? Of course this got me thinking. Do any humans ever reach their full potential? It seems that we never fully open our minds or our hearts. How much more could I grow and bloom myself? How am I stuck or closed? Well, rather than continuing down a potentially melancholic rabbit hole, I decided to learn a bit more about this plant that brought on my pensiveness.

I learned that the turk's cap is in the mallow family, as is the hibiscus. Each flower has a long pistil with attached stamens sticking out of what look like closed, twirled petals. These petals never open up like other flowers. Their twisted appearance looks a bit like a turban, hence the plant's name. They bloom through summer and fall. The plant can grow up to ten feet tall and ten feet wide, requires little maintenance, and, once established, is drought tolerant. All very interesting facts.

But the most enlightening thing I learned about this plant is that its flowers, the ones I first judged as incomplete or not quite right, are absolutely perfect for hummingbirds, bees, butterflies, and other insects. These blooms attract them and feed them. They are a favorite, a superstar, and an important sought-after stop for migrating hummingbirds.

I realized that this plant, which I initially judged as "less than" others, serves a vital role in the lives of other creatures. Now, I am not a complete newbie to this. I know about the complexity of ecosystems and that all life forms on the planet rely on each other. I know that they and we all have different strengths and talents. But, I suppose, in my imperfect human way, I temporarily forgot all of these things in my quick judgment of the plant as not looking exactly as I had wanted or expected. Oops.

I've learned that my life lessons seem to repeat themselves. I forget, and then I learn again. I forget, and then I learn again. This was another lesson for me. A reminder. I do not know the destiny or purpose or life lessons of any other being, plant, animal, or person. So why should I judge them based on my expectations? Or make any assumptions? Or devalue anything or anyone? The answer is, I shouldn't. I'll try to remember this better. On the other hand, I won't beat myself up about it. I'm still not fully open, and I haven't reached my full potential. And that's ok too.

I decide that once I get home, I'll plant a turk's cap mallow in my yard; I'd love to provide another feeding station for hummingbirds. And perhaps it will serve as a reminder for me to accept everyone (including myself) as they are and not as I might think they "should" be.

Big Chair, Little Me

A person's a person, no matter how small.

—Dr. Seuss

Next to a shaded sanctuary area in the garden, there's a spot of whimsy. A giant, bright-blue Adirondack chair welcomes passersby to take a seat. I managed to climb up and get into that chair just once. The experience of sitting in it, with my feet jutting off the edge into the air, and subsequently needing help to get out, made me feel very much like a small child. I suddenly sensed a clearer empathy for children. Often in places or circumstances in which they feel small, they may feel frustrated; they want to climb, to reach, to move, to take action on their own, without help, but can't. Perhaps feeling virtually encompassed by the chair, followed by distinct sensations of struggle in my attempt to exit it, tapped into my body's memory of past similar experiences. It's one thing to imagine how a child feels, another to recall memories of childhood, but a much stronger experience to tap into one's own body's memories, cellular memories.

Back on the ground, I felt a wave of compassion for my younger selves. Pictures of my child self passed unbidden through my mind like a slideshow: grimacing sitting on the kitchen table at night as my mother put countless bobby pins in my hair attempting to achieve a Shirley Temple look the next morning; asking my uncle to lift me up to see my grandmother's ceramic doll figurines that so fascinated me, displayed too high on the wall to see on my own; straightening my back and stretching up my neck in order to see out a car window.

With images of memories in mind, the woman I am now spontaneously sent loving care and comfort back in time to the little girl I was. I felt as if some little piece of hurt found peace.

The chair experience was small, inconsequential, and unplanned. It lasted less than five minutes. But it was powerful. Sometimes life gives us gifts like this. Impromptu. It may seem as inconsequential as a friend offering a stick of gum before she unwraps one for herself. I could easily have missed noticing this gift of enhanced empathy and compassion. I'm sure I miss some like this. We probably all do. Perhaps another gift is the realization that life gives us guidance and opportunities for lessons every day. When we believe this, intend to pay closer attention, and appreciate the lessons, more serendipity, synchronicity, and healing moments will happen.

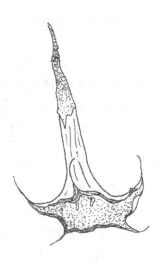

ANGEL'S TRUMPET—*BRUGMANSIA AUREA*

Angel's Trumpet

Well-being is attained by little and little and
nevertheless is no little thing itself.

—Zeno of Citium

Next to the garden's whimsical giant blue Adirondack chair
is a shaded sanctuary garden. Several regular, people-sized
chairs are positioned near a small fountain. It's a place where,
on breezy days, I come to watch the dance of dappled light as
late afternoon sunlight streams through the swaying fronds
of a large areca palm. A variety of trees and flowering shrubs
grace the area. I'm headed there now.

Walking past the bamboo stand, I spot Peggy, the present
resident gardener, on a parallel path. I'd like to thank her for
some tips she gave during the tour I took with her the last
time I was here. She's such a fast walker that I almost have
to run to catch up with her before she's out of sight. We're in
front of the sanctuary garden when I reach her. As I thank
her, I glance into the sanctuary garden. Among the foliage,
one tree stands out. I'm stunned by its appearance.

"Wow, Peggy! Angel's trumpet. It looks perfect!" I exclaim as I dash to the tree. Peggy walks calmly and smiles as I rave.

"The leaves. They're all green. No spots. No holes. No withering. Every leaf looks full of water. Even the bark looks full of life."

The last time I was here, the angel's trumpet did not look well at all. Its leaves had many brown spots. It had no flowers, but that was normal. It was not late summer or fall, when its spectacular pendulant blooms abound, hanging like bells. I had wondered if it would recover from whatever was causing its disease and hoped that, in the fall, I would sit in this sanctuary nook and appreciate its beauty.

This angel's trumpet now exudes health. Its well-being is palpable. I'm deeply moved. I feel a sense of awe witnessing this tree's recovery and display of strength.

"What did you do?" I ask.

"Basically, I helped Mother Nature," Peggy says. "I cleared away the mulch that had gotten packed around the base of the tree, did minor trimming, and gave it the revival food I swear by. That's about it."

"Anything else?"

"Well," Peggy pauses. "This is my favorite place in the whole garden. And I feel a special connection with this angel's trumpet. Maybe that helped too. And something in the tree.

It responded. Improvements started. And little by little it got its health back."

I congratulate Peggy and thank her for her time before she walks on. I'm still marveling at this tree's recovery as I sit down on a chair facing the tree.

I think about the angel's trumpet I have at home. Unfortunately, it resembles how this plant looked when I last saw it. Seeing the restoration of health in this tree, I'm feeling hopeful that my own tree can regain well-being little by little, as this one did.

What about your well-being, Charlene?

I'm hearing this, not in words, but as a sensation in my body. I think about this. What about my well-being? I feel another inner nudge. Hmm... Maybe it could use a little improvement.

The value of well-being is monumental. I know this. I know it from having it and not having it. But no, that's too simplistic and extreme. There are degrees of well-being. It feels wonderful to have it fully. For me that's when smiles and laughter happen easily and often; when I feel comfortable in my own skin; when my thinking is clear; when I feel energized or playful; and when I feel flexible—both in my body and in my life's circumstances. I think, perhaps, mine isn't as robust right now as I'd like it to be.

I remember what Peggy said: "Little by little." Perhaps well-being is a kind of strength, sustained by making many little smart choices day after day. Drinking lots of water, getting enough sleep, and eating nutritious food. Nurturing ourselves spiritually, mentally, and emotionally.

I sigh. I know this. I "get it." But do I "get it" enough? To the extent that choosing the little things that sustain well-being becomes habitual? Every day? I have to admit that it's not yet for me. And I've learned that sometimes well-being diminishes the same way it grows: little by little. And sometimes without notice.

A breeze shifts my attention back to the angel's trumpet and I smile. Witnessing the renewal of this magnificent tree, my commitment to maintain high well-being is again set. I know I can do it, little by little.

Scrambled Eggs with a Side of Compassion

A moment of self-compassion can change
your entire day. A string of such moments
can change the course of your life.

—Christopher K. Germer

While coffee brews, I step outside to feel the temperature
and smell the rain-fragranced air. I heard a downfall during
the night, but it's clear, warm, and dry enough this morning
to enjoy breakfast al fresco. As I clean the tabletop, I glance
toward the tropical plants that act as a privacy screen outside
the lanai. Dappled sunrays beam like spotlights on the croton
varietals. I marvel at the different leaf shapes and the patterns
of red, yellow, and green hues that look as if they were painted
on the leaves, some with a brush, some with a spray, some by
drips and spills. Plump, round rain beads still rest on many
of the leaves. A breeze rustles the leaves and the raindrops
sparkle and shimmer as if to dance in the spotlights. The
scene delights and inspires me. I almost applaud.

Returning to the kitchen, I remember yesterday's encounter with the restored angel's trumpet tree and my commitment to enhance my well-being. I decide to forgo the sugar-topped muffin I plated and scramble two eggs instead.

I think about what well-being means to me. Good health, happiness, a sense of meaning and purpose. And the sense of feeling generally good overall.

I recall the vitality I attained twenty years ago after recovering from advanced non-Hodgkin's lymphoma and its treatment's side effects. The shock of the cancer diagnosis had galvanized me into action. I then created a routine of daily practices I required of myself during the months of chemotherapy and for years afterward. They were simple nurturing things for my body, my mind, and my spirit. I enjoyed them. But there were many of them. It had taken a firm commitment and more than an hour of time each day. But it worked. After that, I cherished the robust health I had regained and thought I would never again take life, health, or well-being for granted. But, little by little, my zestful passion waned. Being imperfectly human, I let this and that gradually fall by the wayside.

Before bringing my scrambled eggs out to the lanai, I top them with fresh parsley, add some bright mixed berries to my plate, and grab a basket of shells to beautify the table. Starting right now, I'm intending to use many little self-care enrichments throughout each day, as I did during my first encounter with cancer in 2000.

The main thing lowering my well-being now is more weight. I gained it little by little over the past few years, after I faced cancer a second time. That time it was breast cancer. The drug prescribed to me after treatment has a side effect of weight gain. Although this drug has been a contributing factor, a more significant one was my own out-of-control eating.

I believe that beauty comes in all shapes and sizes, but, as my body grew, it protested with aches and pains, telling me that this shape and size was not the right fit for me.

Recently I exceeded a certain number and hit an emotional bottom. Then the weight change somehow halted. I haven't begun to release the weight, other than a pound or two, but at least the gaining stopped. Now I'm determined to release the large gain at a healthy rate in healthy ways.

While I sip my coffee, I reflect. I've come to believe that everything in life holds an array of gifts. Last night's raindrops clinging to the croton leaves in this morning's sunshine gave me the gift of delight. What is the gift for me in my weight gain?

One of my grandmothers, my mother's mother, comes to mind. Not quite five feet tall, she weighed almost three hundred pounds. Since she lived with us for many years after my grandfather died in the late 1950s, every day I watched her struggle to do many things, such as climb stairs and get up from a chair.

I felt compassion for my grandmother. I loved her so much. But I also judged her. As a child I even felt embarrassed at times. I questioned why she couldn't lose at least some of the excess weight. Why couldn't she get healthier and be more comfortable?

My grandmother's obesity affected both my mother and me. Whenever either of us gained more than ten or fifteen pounds, we worked to make sure the weight didn't keep rising. I remember my mother often saying she was "watching" her weight. Not dieting, just "watching," as if it were a minor but constant presence in her mind.

I now know what it's like to have trouble getting out of a chair, tying my shoes, climbing up onto my bed, and walking uphill. Sometimes just walking, period. I also know the experience of feeling almost powerless to stop overeating. As if, no matter how much or in what ways I tried, I just couldn't stop, especially at night, watching the news, feeling anxiety, or not being able to sleep. Eating at night got so habitualized that I couldn't fall asleep without eating something, or several things, before getting into bed. Then, I discovered the pleasure of eating something while I was in bed, reading.

Eating while reading before sleep became a routine. A habit. An addiction.

I've realized that my emotional recovery after the double mastectomy was not as easy as my emotional recovery after treating lymphoma with chemotherapy. Not that that was "easy." But once my bald head became covered with baby-fine fuzz and it eventually grew into hair that could be styled in my usual bob, I looked like me again. I felt like me.

On a surface level in my mind, having a double mastectomy, followed by reconstructive surgery, is not such a terrible thing. One can rationalize that the cancer is gone and be appreciative of that, which I did. I even developed a comedic attitude, joking about getting "a perky lift" with Medicare paying for it. But it truly was a trauma. On deeper levels, my body was in grief. The implants were uncomfortable. The physical healing process seemed slow. I didn't look or feel like the me I knew. So I ate for comfort.

I suspect my grandmother ate for comfort too. She was likely depressed. I know she had been unhappy in an arranged marriage to an older man she didn't know well. Although she never said it, I think she neither loved him nor liked him very much.

I hardly knew that grandfather. I saw him infrequently and he died when I was only ten. When I did see him, he paid little attention to me. I remember three things about him from visits to their walk-up apartment in Hoboken. The first is seeing him come home for lunch from his work at a bakery with a scowl on his face and a loaf of fresh bread under his arm, expecting a bowl of hot soup to be ready for him on the table. The second thing happened shortly after I got my long hair cut into the then-popular pixie style. I was nine. I thought it looked cute. I thought I looked cute. But I heard him comment to my mom, "She looks like she was run over by a lawn mower." The third thing I remember is positive, in part. He had two parakeets. I could tell that he loved them by the lilt of his voice when he talked to them in Polish. That showed me a likeable side of him. However, he seemed kinder and more attentive to them than he was to me, to my mom, or to his wife. In the way of a young child, I sensed something sad and empty in their home and I felt a caring protectiveness for my grandmother.

My grandmother had a fatal stroke in 1974. All these years later, in this numinous moment, compassion for her fills my whole heart, not just a part of it like it did when I was a child, because now I can empathize, not just sympathize. I understand, not just in my imagination, but from experience. I get it.

My recent struggles with weight, along with the process of aging, have moved me closer toward truly accepting people as they are. Who am I to judge my grandmother—or anyone— who is overweight or underweight or has any issues with eating? Or any "issues" at all for that matter? It occurs to me that, although judgments are not detected when we step onto a scale, they do weigh us down. I don't want the burden of them. Nor do I want to burden others with them.

I bring my breakfast dishes into the kitchen. While washing them I resolve to let go of judgments, as much as I can and to give myself and others regular doses of compassion.

But, wait. I am forgiving myself for mistakes, for being weak, for missing the mark. Doesn't this form of self-forgiveness perpetuate self-judging? What I need to do, what I shall do, is forgive myself for judging myself. I forgive myself for judging myself as weak. That feels right. I may have weak moments. I may have experienced weakness. But that doesn't mean I am weak. The truth is: I am strong.

I intend to replace "not good enough" thoughts with affirmations of worthiness and with embracing myself just as I am.

I wish we would all forgive ourselves for all the times we thought of ourselves as not strong enough, not good enough, not thin enough, not smart enough, not pretty enough, or just not enough in any way. Enough of that. We are enough.

True forgiveness. Sweet compassion. Unconditional love. It's what we all deserve.

YESTERDAY, TODAY, AND TOMORROW—
BRUNFELSIA PAUCIFLORA

Yesterday, Today, and Tomorrow

The past is behind, learn from it.
The future is ahead, prepare for it.
The present is here, live it.

—Thomas S. Monson

It's the start of a new year and my first Sanibel stay since early autumn. I'm eager to see what's blooming in the garden now, especially because I'm rarely here in January. And since I arrived last night at dusk, by the time I unpacked my car, it was too dark to see the garden.

I've signed up for a watercolor class. I understand we're to paint something that's in the garden, with each student picking her own subject after a tour, which is about to begin. Our instructor, Mary Lou, leads the way. Her love for this garden and joy from painting are indisputable and contagious. As she calls our attention to details of color, shapes, textures, light, and shadows, and to the interplay among objects, I'm captivated.

Over the years I've taken countless tours of this garden, mostly on my own but also in groups led by the then-resident gardener. Tours reflect the views of the guide, so I've always come away enlightened with new information and enriched by how it's shared. Mary Lou's perspective as an artist is deepening my gratitude to be in the garden.

All the students seem to know what they'll be painting. I haven't yet felt a pull in my heart toward a particular plant, tree, leaf, or flower. I wait for that to happen as we continue through the garden.

It happens as we approach an area of shrubs covered with small flowers in various shades of purple, blue, and white. Like iron to a magnet I'm drawn to the delicate blooms. According to the signage under the shrubs, the proper name for this plant is *Brunfelsia pauciflora*. Its common name is yesterday-today-and-tomorrow.

"It's also known as past-present-and-future," says Mary Lou. "Both common names are fitting because each bloom generally lasts just three days." She goes on to explain that as each flower lives its life span, it changes from a vibrant periwinkle to a medium blue-violet, and then fades to white before falling to the ground.

Although I'm interested in acquiring more horticultural information about this plant, right now I'm eager to paint. After completing the tour, my classmates and I scatter.

One heads toward a bottle tree, another to a ponytail palm, another to a yellow hibiscus. Others are off hither and yon, while I return to the *Brunfelsia pauciflora* shrubs.

Before sketching, I observe the shrubs overall and notice that vitality and beauty are maintained in each stage of each flower's life. The elder, white-petaled flowers appear as healthy as the young, deep purple ones. Also, the spread of colors on each shrub suggests that blooms of all ages are mixing and mingling harmoniously. I smile at the symbolism of that. A particularly graceful stem with one dark, one medium, and one lightly hued flower attracts my focus. I give it my full attention as I relax my hand and make faint lines on my sketch pad to indicate the flowers' edges and the petals' curves. After sketching, I wet the paints and mix blues and purples on my palette until I have a spectrum of periwinkle hues. I hope my painting will convey the delicacy I see and the gentle loving feelings these flowers evoke in me.

As I add final touches, I perceive something in these three flowers that is deeper than their outer beauty. And it evokes thoughts and feelings beyond my affectionate appreciation for their tenderness. I see some ways we are like them and they are like us.

Each particular yesterday-today-and-tomorrow flower can be recognized as a unique botanic being. This may seem like an unusual way of perceiving a flower, but, like every human, each individual flower is a one of a kind, once-in-all-eternity creation. Original. As is every life form. Recognizing this about ourselves and others—humans, plants, and animals— can stir our inborn awe and reverence and elevate our treatment of all beings.

Like *Brunfelsia pauciflora* flowers, we have a past, a present, and a future. But every day is always today. We, like the flowers, are always only in the now. This moment. However, unlike flowers (as far as we know), we possess the awesome power of time travel. Our consciousness enables memories to transport us back to yesterdays and imagination to carry us into tomorrows. What amazing gifts! How much do we appreciate them? Do we know how to use our memories and our imagination to our best advantage?

We can gain wisdom from the past when we look back with love, compassion, and readiness to learn. We can prepare for the future by using imagination not to visualize scenarios that scare us but to envision possibilities that bring forth the best in us. We can accomplish this when we bring our time travel gains into the only time and place where we can act upon them. It's where we live. It's the here and now.

Aging Like Brunfelsia

You don't stop laughing when you grow old,
you grow old when you stop laughing.

—George Bernard Shaw

After breakfast, I walked through the garden to its north edge, where it meets boat docks alongside a canal. I'm sitting in a dock chair now, facing the water, with many trees and shrubs behind me. I'm appreciating the broad sea grape leaves and their twisty trunks and woven branches, but I'll write about them another day.

I've been thinking about the *Brunfelsia pauciflora* flowers, better known as yesterday-today-and-tomorrow flowers, that I painted yesterday. I'm intrigued with the white ones, the elders, the old-age blooms. Even though they are in the last stage of their three-day life span, they look as vitally alive as the deep purple ones. The elders may not be as vibrant in color as the *Brunfelsia* youth, but they seem every bit as sprightly, hearty, and hale.

How do they manage to maintain their zest, this human elder wonders and imagines. What might we learn from them?

Maybe they are happy. Maybe plants have the equivalent of emotions, and the aged flowers are just as joyful as the young. Maybe, in some yet-to-be-discovered mysterious plant behavior that we can't perceive, they even laugh. I imagine them laughing often, the moment their buds open to the light, as they grow through youth, into middle age, and on to old age. Maybe all flowers laugh. Or maybe there is some truth in what Ralph Waldo Emerson wrote in his poem, "Hamatreya," that the "Earth laughs in flowers."

Who can know or say with certainty that flowers are not the earth's laughter? Mother Nature's laughter? God's laughter? Even if we dismiss the literal possibility, picturing Emerson's poetic expression and being amused or enthralled by the image can help keep us young at heart.

George Bernard Shaw declared, "We don't stop playing because we grow old; we grow old because we stop playing." He was right: both laughter and play keep us young at heart. From my perspective and experience, seeing the world with wonder keeps us young too.

I saw this in my father's parents. Although they looked old to me when I was a young child, I didn't think of them as old because they were young at heart. They liked to be funny, especially on Sundays when our whole extended family gathered for dinner. After dinner, when all the women and kids were in the kitchen, my grandfather would pretend to sneak in to steal a kiss from my grandmother. Since it was a ritual, it was no surprise to my grandmother, but we kids fell for it. He'd tiptoe in, tap her on the shoulder, and try to plant a big kiss on her cheek when she turned. Then he'd run off laughing as she tried to swat him with a dish towel. For one of their anniversaries, he gave her a plush skunk toy with a card that read, "Thank you for putting up with me, Anna. Your skunk, John."

The tables were turned every year on April Fools' Day when she played a trick on him. Every year he fell for some silly prank she created. April 1 was her birthday; I think his playing along was part of his birthday gift to her.

I hope the gifts of love, laughter, playfulness, and joy my grandparents gave each other ripple on as I share stories about them and try to imitate them in my family.

I also try to follow the advice and example given by my aunt Angie. When I was a teenager, we both worked as "directory assistance" operators at the phone company. Sometimes after a shift we'd stop at an ice cream parlor and talk about important things. I remember one time she said, "I love this job, Charlene. If I hated it, I would quit. Don't stick with anything you don't like. Stick with what makes you happy." In one form or another, she often said that. "Do things that make you happy. Choose things that make you happy. Be grateful and you will be happy," she would say. "To live long and stay young, be happy. And take a nap every afternoon." Aunt Angie walked her talk, being happy, napping, and living with gratitude to her ninety-sixth birthday. To the end, she was generous with smiles and encouraging words. Every time I phoned her, she said she was happy and asked me if I was.

Aging well is a significant topic for me now, as I suppose it is for about seventy-five million of us baby boomers. I never minded a birthday or turning thirty, forty, fifty, or even sixty. I felt youthful inside. Although I may have thought sixty was old when I was thirty, I certainly did not think it was old on my sixtieth birthday. Being healthy had a lot to do with it. At sixty I could look back and see good things in my rearview mirror, including triumphing over cancer. I celebrated sixty with a simulated skydive.

But seventy feels quite different. Maybe it's because of the added weight I have been carrying. There is also the reality that no matter how well I survive and thrive into old age, there are fewer years ahead than there are years behind me. Of course this was true at age sixty as well. I suppose I'm becoming more acutely aware of it as I approach the ages at which my parents passed away. My mom was seventy-one when she succumbed to the leukemia she had kept at bay for three years. My dad died suddenly at seventy-four from a heart issue. I have peace knowing that his last day had been a happy and playful one. On a Florida vacation with his jokester brothers, the day included fishing, laughter, balmy weather, and good food. He was savoring an ice cream soda with his brothers one minute, before clutching his chest the next, and falling to the floor a moment later.

I continue to be surprised and somewhat envious when someone my age talks about a still-living parent. A few friends even have both of their parents in their lives. I'm not resentful. I don't begrudge them. What I feel is a combination of shock and longing. The shock is realizing that, had they not died in their early seventies, my parents would be in their nineties. The longing is a yearning to have had them in my life during the past twenty-something years. I suppose I am even a bit jealous of my cousins who had their mom, my aunt Angie, in their lives for decades longer than I enjoyed my parents' presence. The painful twinge of missing them doesn't last long, however. They lived well. I didn't have any serious unresolved issues with them. I do wish I had asked more questions though and learned more about their early lives.

This line of thinking leads me back to thoughts of my own aging and to the universal question: How does one age well?

I think we've all seen characteristics of negative aging in such things as bitter complaining, blaming, faultfinding, rigid thinking, and excessive focusing on health problems. There's often a lack of humor and joy too.

Positive aging, on the other hand, is characterized by such things as an optimistic attitude, caring about others, appreciating nature, and an interest in ongoing learning. There's also often a readiness to overlook faults and to forgive. Studies have determined that we've all got a mixture of characteristics, some of which tend to promote positive aging, some negative aging. But regardless of our personality traits, we each have the ability to consciously choose how we respond to life. Responding with courage, compassion, wisdom, hope, and humor can help us age well.

I've heard that there used to be a sign in Alaska at the start of a rutted unpaved road that went more than fifty miles. The sign read: "Choose your rut carefully. You are going to be in it for a long time." I don't know if it is true or not, but it makes a good point.

At any age, it makes sense to ask, "Do I like the course I'm on? Do I like where the patterns in my life are taking me? Or do I want to get into a different groove?"

I like the groove I'm in, but I try to look with care at the choices I'm making each day, to be certain that I'm fostering the qualities that promote positive living and aging, as my parents did. They both demonstrated a great deal of hope, humor, kindness, and faith.

When my mom was diagnosed with leukemia, despite a grim prognosis, she hoped and prayed for a cure. For three years she was able to live quite normally and well.

Then her condition changed. Decline was rapid. And my mom began to ask a certain question in her daily prayers. It was never, "Why me?" Realizing there was nothing more she could do, she asked, "What would you have me learn today, God?" She asked this question every day for the remainder of her life, which was less than a few months. And she got answers. Not every day, but often. During that time, my mom quite often said things like, "Oh" and "I see" as her eyes brightened.

I was with her much of the time toward the end of her life. Several times, with a knowing smile, while moving her hands as if she were putting pieces into a puzzle, my mom would say, "Everything is falling into place, Charlene." She didn't share what she learned with me. But I got the sense that she felt incomplete issues of her life being resolved and saw meaning behind previous confusions and hurts. She may not have received a cure, but she achieved a deep healing. Despite the weak, frail, and ravaged appearance of her body, she exuded strength, beauty, and joy. I saw her peace.

A few weeks before she died, my mom was admitted to the hospital for a blood transfusion and to be treated for an infection. After the procedures, we took a slow, short walk. With one hand resting on my arm, the other pushing her IV pole, my mom made her way around the corridor. She stopped in front of each door. If a patient was awake and made eye contact, my mom smiled and said hello. If the patient was sleeping, she gestured a little sign-of-the-cross blessing. We stopped also at the nurses' station, where my mom thanked everyone for their care. I saw in her the beauty, grace, and vibrancy of the white *Brunfelsia* flower.

Holy Mystery

When the veil is torn apart and our vision is clear, there emerges the recognition that all life is connected—a truth not only revealed by modern science but resonant with ancient mystics. We are all one, connected and contained in a Holy Mystery about which, in all its ineffability, we cannot be indifferent.

—Judy Cannato

Other than the bees, butterflies, and bunnies I've been watching, I seem to be alone, at least for this moment in this vicinity of the garden. I'm sitting under a little white wedding arch. There's not another human being in sight. That will soon change, I'm sure. A lot happens here. A guest with an armful of used towels may take this path to the linen exchange. A child may skip by and smile on her way to the pool. The handyman may pass with tools to fix something. But human activities won't disturb the resplendence of nature here. Like the bunnies, the bees, and the butterflies, like the box turtle I saw earlier, like me, they—we—are all part of this garden habitat. I want to simply sit in gratitude for the beauty that surrounds me.

A lot has happened in this garden, on this island, and in our lives since we first crossed the causeway and stayed in this resort decades ago. A powerful transformation has taken place here. Over the years, a beautifully landscaped resort became a garden environment. That garden environment was upgraded to a botanical garden. While still holding its status and certification as a botanical garden, this place has been elevated to a garden habitat. It's a subtle but sublime change. This garden is not just here to please people (which is a fine goal, a meaningful accomplishment, and a worthy service). A habitat is a place that supports and sustains life, the life of all the plants, animals, and people who live there or visit. The life of everything is honored. The needs of everything are considered. The well-being of everything is supported and nurtured.

I sense a Holy Mystery here. I don't claim to have had the veil lifted or torn open. I have not had a vision or a mystical epiphany. But the curtain between what human eyes can see and the splendor of life seems thinner, more opaque here now.

I wonder if it's because this garden is cultivated, fostered, and loved as a habitat. The present resident gardener defers to the needs and wishes of the plants when she grooms. In her approach, she aims for controlled wildness. She told me that earlier today when I said I'd love to see all the vegetation be allowed to thicken and grow even taller than it is. "I'd like it to look and feel like Costa Rica," I had said, jokingly explaining that I had not yet actually been to Costa Rica. We laughed together as she said she understood.

A butterfly flits by the arch. I hear the coo of a bird nearby as I close my eyes and take a deep breath.

I could be imagining it, but I'm getting a sense of the garden experiencing freedom and joy. I'm aware of the plants. Are they aware of me? Can they sense my feelings about them? I wonder. Might plants, like animals and young children, though without speech, know when they are authentically loved? And appreciated? Maybe that is how a green thumb works. Maybe plants respond to emotions as well as physical conditions. And do they communicate? I can almost hear a language spoken among the plants. Not with my ears of course, but as a humming throughout the garden.

How vibrantly alive this lush habitat is—the plants and all the creatures who exchange air with them. How wonderfully oxygenated it is. It seems almost as if I can feel and hear the exhalations of the trees. I wonder how many of us ever think about the vital interdependency we have with all living things. We need each other to thrive, even to survive. As I observe, listen, and feel, I perceive the oneness of us. I take another deep breath. I can't help but be in love and be in joy here.

SHAMPOO GINGER—*Zingiber zerumbet*

Talentry

Everybody is talented because everybody
who is human has something to express.

—Brenda Ueland

Shampoo ginger grows right outside our door. Next to it
stands the miracle berry. Simpson's stopper is across the
street. These are among the plants recently showcased for
their special talents on the garden tour. Shampoo ginger plant
is so named because, when squeezed, it oozes a substance that
can clean hair. Simpson's stopper stops diarrhea. Miracle
berry has a low sugar content, but when chewed removes
bitterness from the next food being eaten. It's been used to
help people taking certain sour medicines to stomach them
better. The fruit has also been shown to help many cancer
patients being treated with radiation or chemotherapy to be
able to taste and enjoy the flavor of foods again.

We may not normally refer to plants as having talents; we are
used to thinking in terms of their usefulness. And, yes, they
do contribute enormously to human wellness, and that is how
I describe them. But there is more to it as I see it; each plant
has special abilities and talents. Like us, each plant can make

a variety of contributions to its environment, to animals, to humans, and even to other plants. Besides the essential gift of oxygen, they provide much more. Of course they provide medicines and food. We don't use as many plants medicinally as we once did, but, for the majority of human existence, the garden of earth served as humanity's pharmacy. And we used to eat a much larger variety of plants. Many more plants are edible than people realize, even in this garden. The number of plants in the typical modern human diet is much smaller and much less varied than it used to be at certain times in history and prehistory. Shade, temperature regulation, soil erosion control, and carbon sequestration are all additional talents of plants.

This may be a different way of thinking about talent for some people. So often we hear of people being talented in specific areas only, such as in athletics, academics, or arts. Those are valuable talents, but there are so many more abilities and skills that are not recognized as such. Talents need not be extraordinary abilities. Everyone has a mix of skills. Think of your special interests. What do you like to do? What are you good at?

During 1999 and 2000, I visited classrooms in a variety of school settings throughout America, where I read my book *The Twelve Gifts of Birth*. I then engaged the children in discussion about their inner resources including the gift of talent. The talents mentioned at first were typically athletics, academics, or arts. But after more discussion, the additional talents mentioned were more varied. The ability to cheer people up, taking good care of younger siblings and pets, solving puzzles, memorizing, finding things that are hidden, and making people laugh were among the talents claimed. I believe that enlarging our definition of talent can enhance our learning ability, performance, and happiness and lead to increased self-worth and greater success. I encourage you to stretch your thinking about your own talents.

Blessed

Being blessed is a condition of the
heart and a frame of mind.

—Errin Rhorie

We're driving to Sanibel, Frank and I. From this turn, we
have just six miles to go. I'm feeling eager yet reluctant to
see the island after Hurricane Irma battered it last month.
Like everyone who had a stake of any kind on Sanibel, Frank
and I had come to terms with the possibility that all might
be lost. A twelve-to-fifteen-foot storm surge was predicted;
Sanibel could have been claimed by the sea if that surge had
happened. Fortunately, a tsunami-like water rise didn't
happen. The tides and other conditions were in Sanibel's
favor. We're feeling lucky and grateful.

My gratitude is tempered by grief and survivor guilt. The grief
is due to wind damage. We've learned that thirty-five trees
within our development are either gone or severely damaged,
including several Hong Kong orchids, an African tulip tree,
and a large frangipani. The shaving brush tree and a wax
jambu are in critical condition.

We're at the causeway tollbooth, in the SunPass lane. The moment we're on the other side, we open all the car windows. It's a ritual, started the first time we crossed this causeway decades ago with our then young daughters.

As if the storm never happened, the causeway offers its usual delights. Sunlight sparkles on the aquamarine water. A pelican plunges in for a fish he spotted.

We appreciate the beauty and the breeze. But our excitement is subdued. We're concerned for Sanibel's lush foliage. And our relief about making it through Irma without devastation is moderated by sadness at the ruin that occurred on Puerto Rico, St. John, Barbuda, and other islands.

I ponder. How do we balance gladness about our good fortune with sympathy for the misfortune of others? It happens throughout life. Sometimes in reverse. We experience misfortune when others have good luck. How do we maintain equanimity through all of life's losses and gains?

We roll off the bridge onto Causeway Boulevard.

Chopped limbs, branches, and other debris are piled high on both sides of the road. When we come to the four-way stop at Periwinkle, I notice serious expressions on the faces of drivers and pedestrians. Not quite somber, but serious. No smiles, but a courteous acknowledgement of one another. There seems to be a softness, a gentle common understanding.

It's one of the positive things that can happen as a result of disasters or suffering.

Frank and I remain silent during the short drive down Lindgren Blvd. and the left turn onto East Gulf Drive. Our destination is in sight.

My stomach tightens when we reach the Moorings and I see that at least half of the vegetation that screened the front of the property on the canal side of the street is gone. Briefly shutting my eyes, I remember to lean into hope and take a deep breath. Can I shift to see what remains instead of what's missing? Can I become open to all that I see and feel instead of resisting it?

Looking again, I almost laugh because of a spontaneous leap in my imagination. There's humor and hope in what I now see. Pointing, I say, "Those gaps in the plants remind me of a child's missing teeth." Frank chuckles. Just as permanent teeth will naturally come in to fill a smile, new growth will fill the holes here, I trust.

Turning right into our driveway, we see that the Hong Kong orchid tree is indeed missing. Parking and stepping out of the car, Frank observes, "The royal poinciana is still here. Magnificent. The staghorn fern's still hanging from it, too. Good."

I sigh in relief, thankful that the dense greenery remains around our small screened lanai also.

So much is here. Yes, much is lost. But more is here. In turns I feel gratitude then grief. I wonder what the individual plants are experiencing. Might they think and feel in their own way? Maybe they too are grieving for the plants that succumbed to the storm yet are glad to have survived themselves. I feel a surge of love for all the life in this garden.

Before helping to unload the car, I pause. I give my full attention to what I'm seeing and sensing right here, right now. A word comes to mind: blessed.

That's what we are. Beyond lucky. As well as being grateful, I'm feeling blessed. Ah, this may be it. The answer. At least part of how we maintain peace and well-being through life's gains and losses. Noticing the good, finding the gifts, and recognizing the blessings.

Post-Op

When the Japanese mend broken objects, they aggrandize the damage by filling the cracks with gold. They believe that when something's suffered damage and has a history, it becomes more beautiful.

—Barbara Bloom

The car's unpacked, the clothes are put away, and the refrigerator is stocked. Whew. I flop on the couch for a few moments' rest while Frank changes his clothes for a walk. I have to admit I feel beat. It's five weeks after Irma, and we're here for five days. Besides checking on the health of the garden, I'm here to heal myself. Before driving here this morning, I had a five-week check-up.

A year ago, I had a double mastectomy after cancer was found in both breasts. But five weeks ago, I had to have the reconstructive implants removed and replaced due to a bacterial infection. The infection is gone. Although physical healing is happening quite quickly, emotional healing is slower.

As we leave the condo and turn onto the path, we see many changes to the garden since our last visit. It's like a post-operative unit in a hospital. Many plants were surgically treated, their irreparably broken boughs and branches cut and removed. Plants beyond help are gone. But some are tied, supported with ropes and tape like casts and bandages. It's evident that all are supported now with very caring attention. I wonder, does this garden know that it is well loved and cared for? I hope so. I think it does. I think all life forms "know" something deeper than we imagine they do. Something about life, its force and resilience.

I empathize with the trees whose diseased and damaged limbs were removed for their survival. Like this garden, I experienced a trauma. I'm still adjusting to the incisions and scars and changes in my own body. I'm grieved by the losses.

But as we walk farther through this garden, I'm also encouraged.

It's early fall, a time when garden growth normally slows. Yet, near each path we follow, I see light green. A lot of it. It's that color of fresh new growth—a harbinger of spring. It's pushing up through the soil. It's emerging from the tips of bushes and trees, especially the ones that had major surgery. All around me at every turn I sense eagerness to grow, a "will to live," and tenacious strength and determination. I'm reminded that the urge to thrive is innate in all of us.

At times, in order to survive, grow, or thrive, we living things need to be "pruned." Sometimes we have parts of ourselves cut back or amputated. Sometimes, for the good of all concerned, we need to sever a relationship, uproot a fear, alter a limiting belief, or drop an unhealthy habit. But this clears the way for renewal.

The bright, light, young green I see on so many trees and shrubs leads me to hope. Refreshed, I turn another corner to explore more of this garden, still very much alive.

Mamey Sapote

Once we believe in ourselves, we can risk
curiosity, wonder, spontaneous delight, or any
experience that reveals the human spirit.

—E.E. Cummings

I wake to the smell of coffee and the sight of slivers of sunlight streaming through slits in the window blinds. It's a calm, clear morning, so different than the days when Hurricane Irma hovered and howled here. I intend to put thoughts of the hurricane behind me. I've come to terms with losses, and I'm confident that the garden will recover in the care of Mother Nature and the hands of caring garden workers.

After a few stretches, I sit up, ready to enjoy this new day.

Stuffed into one of my slippers is a note from Frank. "Good morning. Gone for a beach walk. Brought my metal detector. Maybe I'll find a gold doubloon. Ha ha."

After a cup of coffee, I grab an apple and head out for just a short walk. Beach or garden? The beach can wait, I decide. We'll spend a good portion of the day there later.

I follow the closest path. At the next crossway, I notice mamey sapote. How had I not noticed this vibrant young fruit tree yesterday? I've taken a special interest in this tree. I remember the name "mamey" and think of this tree as female because I had a great aunt Mamie. Even though this mamey tree is pronounced *mom-may*, not *may-me*, I stick with "Mamie."

Today, instead of reminding me of Great Aunt Mamie, the stance of this tree shows me youthful cockiness. Mamey trees remind me of an adolescent who has not yet recognized their vulnerability. Even after being battered by the winds of Irma, the tree stands ramrod straight, with arm-like branches held high, as if in a dare.

But a closer look suggests that the mamey's arms are raised in praise, in prayer, and in welcoming whatever experiences are on their way. In welcoming visitors to this garden, too.

From this perspective I get the sense that the mamey has trust in Life's Maker, in Mother Nature, and in where she is rooted. In her welcoming stance, she also seems ready to give hugs and ready to share her fruit, asking for nothing in return.

If, as I've heard, what we notice in others is a projection and reflection of what is in us, I appreciate noticing the mamey's qualities.

I guess I'll have to own up to the cockiness I projected too. During my adolescent years I did feel pretty invincible. Far past adolescence, I now know that we're all both strong and vulnerable.

I realize that I'm hungry and think Frank probably is too. I say goodbye to my tree friend, mamey, turn around, and think of breakfast. I wonder what goodies Frank uncovered and experienced on his walk.

WEEPING HIBISCUS—*HIBISCUS SPP*

Bella Hope

After every storm the sun will smile; for every
problem there is a solution, and the soul's
indefeasible duty is to be of good cheer.

—William Alger

We're still at the Moorings on our first visit after Hurricane
Irma. For the past few days, I've been reflecting and
journaling in various locations in the garden.

I've just visited every bush and shrub in the hybrid hibiscus
area. Weeping hibiscus and life saver appear to be the only
varieties in bloom now; that seems highly symbolic to me
after the storm. But perhaps I missed a flower somewhere.
Maybe a bud of another variety is just ready to open. I'll look
again. I'd especially like to find the hibiscus with my favorite
name, bella hope.

Bella hope. Bella hope. Where are you? I'm calling
in my mind.

I look everywhere for bella hope. She is not here. Her
identification sign isn't either. The sign and the plant may

be gone, but I believe that beautiful hope is still alive and well in this garden.

I see it in the hundreds of little seedlings that have popped up below the nearby royal poinciana tree. I imagine that during the hurricane, the seed pods on this tree were blown off by wind. Heavy rain brought them to the ground, pounded them open, and soaked them thoroughly. From those beaten seeds, a luxurious blanket of new sprouts has started. I see in them the urge, the will, the strength, and the courage to survive. And instinctive hope that they can.

Hope. What is it, exactly? We learn the word at an early age as we begin to make wishes and hope they will come true. If we feel optimistic about receiving what we want, we are said to have hope. But at deeper levels, hope is much more than making wishes, imagining a good outcome, and experiencing a sense of promise. It's a profound place, where, instead of despairing, we trust that life is good even when something unimaginable happens. Hope allows us to live in peace with what is and what will be while we adjust to a new normal and await direction from inner wisdom. Hope offers fresh starts and helps us find a door or a window through which we can move forward from a dispirited place to a meaningful future.

I walk and search a bit more, looking for even one hibiscus bud that's ready to open, but I find no other blooms or buds. I return to the bush loaded with pink, downward-hanging weeping hibiscus flowers. Then I visit the one filled with red, upward-looking life saver flowers. There is such a contrast between these two.

To me, the weeping hibiscus and life saver represent sadness and grief over what has been lost, what has been damaged, and what still hurts. But, at the same time, there is also gladness and gratitude about what has been saved and that the storm has passed. In the upward-looking life saver, I see hope for the future. Hope that everything here will recover and bloom again.

Before I move on, I'll delight in the names of the varieties here. Like wearing a name tag at a party, every plant in the garden "wears" its name on a sign. And it is party-like in this section of the garden, particularly when many blooms are in attendance and it's a riot of color. It's a riot of names too. In every tour I have taken here, someone has commented on the great names of the hibiscus varieties. The names alone elicit smiles.

Whistling Dixie, Misbehaven, Tarantel, Fire Eater, Bold Idea, Holly Pride, Crawfish Pie, Bloom Blazes, Chariots of Fire, Five O'Clock Shadow, Voodoo Queens, Standing Ovation, St. Elmo's Fire, Top Gun, Connie's Surprise...

Such fun, lively names. There are others, too. And perhaps some new ones will be added. At times there have been more than 250 varieties here. I trust that those now here will all be back in bloom. As it is said, "For everything there is a season." Their seasons will come again.

In addition to hope, I sense joy in this garden, always, even when little is blooming. Unlike happiness, which is dependent upon outer conditions, joy is a constant in us too. It is always present.

Gratitude is a key to open joy. Joyfulness does not lead to gratitude; gratitude leads to joyfulness. I believe this is true because I've seen it in others and experienced it so many times myself. In sad times, the joy that is experienced may not be standing-ovation joy; perhaps it's more like answered-prayers joy or divine-grace joy. A parking space by the door when it's raining, the smell of fresh cut grass, clean sheets during a time of illness, or the first sight of daffodils in spring. Gratitude for small things, tiny things, can lead to joy. Simple-pleasure joy. I'm feeling it in this garden now.

What My Toes Find

People usually consider walking on water
or in thin air a miracle. But I think the
real miracle is to walk on earth.

—Thich Nhat Hanh

Frank and I are in the warm gulf water at low tide, beyond the exposed sandbar. A long hill of accumulated sand that runs along the shoreline, the sandbar creates a lovely trail with a warm tide pool next to it on the shore side. Birds dive for small fish in the tide pool and beachgoers stoop to examine shells.

On the gulf side, the water is shallow for quite a distance. In chest-high water, we swim, float, talk, and walk, when my big toe bumps against something hard and round with a sharp edge. I suspect it's a large piece of a broken shell. Finding shells of all kinds is not uncommon here, in one of the shell capitals of the world. But something about this feels unlike any other shell I've found. With curled toes I lift it up and bring it to where I can reach it with my hand.

I'm taken aback at first, not immediately recognizing it. It's about three inches across and soft, almost furry. It's covered with tiny purple spines, like short hair.

It's a live sand dollar. I'm used to seeing these as white shells, the skeletons of this species. This living creature covered with what looks like purple hair captivates me. It feels like a privilege to hold it. After Frank looks closely, he's captivated too. We take turns holding it, and then gently return it to the seabed. We talk about how lucky we feel and how wondrous that was.

Frank's expression shows that his toe has touched another one. We begin to dig about a little with our toes. We discover that everywhere we dig, they are there, just under the sand, in layers. We have no idea how deep or how far in any direction, but it seems like thousands of living sand dollars are safely hidden right beneath our feet.

Even though they seem well protected, we walk with care. We wonder how far they go, so for a while we let our toes explore. Yup, they're this far… And this far… And this far…

That they are plentiful and safe beneath our feet is thrilling, especially in light of the fact that the numbers of so many species are dwindling. We wonder how often live sand dollars have been just under our feet without our knowing it. We feel as if we've discovered a great treasure. But of course we don't claim it; the joy of the discovery is what we keep.

A Hundred Million Miracles

Miracles happen every day. Change your perception of
what a miracle is, and you'll see them all around you.

—Jon Bon Jovi

It's hours later, well past sunset. Upon returning to our condo
from the beach after our sand dollar experience, we do a
little research on these fascinating creatures. We learn that
sand dollars are a flat form of sea urchin that can be blue,
purple, green, brown, or black. They use their thousands
of tiny spines like legs to move around the seafloor and
burrow under the shifting sand. They live six to ten years
in the wild, and their age can be read by growth rings on
their exoskeletons, like trees. They frequently gather in large
groups, and there can be more than six hundred of them in
one square meter!

I'm still feeling amazement thinking about the sand dollars
and our "discovery" of them. There must be thousands of
them where we were today. I wonder how many are on the
planet. A million? Might there be a billion?

Their presence feels like a miracle to me. Some, maybe many, people would not call finding a living sand dollar and then a profusion of them, or the sand dollar itself, a miracle. We have different definitions of "miracle." And I am guessing that most of us live our lives giving little thought or regard to the miracles all around us.

I remember a quote I've heard, supposedly attributable to Einstein, that says, "There are only two ways to live your life: as though nothing is a miracle or as though everything is a miracle." I am sure that I've been at both ends of that. Some days I do live in a routine or blasé way. But on some days I feel like singing the song about "a hundred million miracles." The sand dollars have reminded me that if I'm not noticing miracles, I shouldn't assume they are not there. When I'm willing to see with eyes of wonder, I might find one right under my feet.

The Generosity of Trees

There are those that give with joy,
and that joy is their reward.

—Kahlil Gibran

Taking a short break from the beach, I'm sitting on a bench under large overhanging branches. This has me contemplating and feeling thankful for trees.

A canopy of comforting, cooling shade is a simple but very welcome gift from these trees after hours in the hot sun. The delight felt from watching the dance of dappled light and shadow through branches is another. Yet another is the peaceful feeling evoked by the soft sound of rustling leaves. These gifts come effortlessly, purely from the trees' presence. The trees don't have to work at giving these gifts. But they can be so appreciated.

For many birds and other creatures, the trees here provide safe housing. Plants of all kinds clean the air and prevent soil erosion everywhere. Some may even provide healing or comfort to people through essential oils, teas, or herbal treatments.

Every day, trees are giving the air and thereby all creatures an abundant supply of oxygen. How often do we remember and appreciate this fine breathing relationship we have with trees and with all plants? It's a miraculous symbiosis. We literally couldn't live without plants.

Whenever a tree is bearing fruit here, the fruit is shared. Often there are baskets of it set out for guests to take freely. At various times, I've enjoyed bananas, avocados, mangoes, star fruit, jackfruit, and more. Yes, one might argue that this sharing is done by the humans here and not by the trees, but consider how the trees give of their fruit freely without asking for anything in return. Of course, one might suggest that it is part of the workings of nature, that they "give" their fruit expecting that offspring will develop from them. That may be, but it seems to me that they give easily, without worry.

Trees and plants do not ask for payment. They do not negotiate for or expect anything in return. They don't wait for the highest bidder. They freely bestow their blessings.

We're programmed for survival. Whether through difficult personal experiences or from family upbringing, over time, many of us have learned to fear loss or lack. For some of us, it's even become a habit, an automatic response.

Trees have become a touchstone for me—a reminder to have the courage to give. Of my time, my talent, my resources, my emotional support, my love... I aspire to this. To give and keep giving, with faith that everything will be ok. Nope, it's not always easy. But I aim to give as the trees give, without stress, without worry, without fear. To give instead with faith and joy.

Never Say Never

Never is a long, undependable time, and life is too full of rich possibilities to have restrictions placed upon it.

—Gloria Swanson

We almost closed our hearts and turned our backs on Sanibel in 2004, when we saw the island a month after Hurricane Charley pummeled it.

Anyone who visited Sanibel before August 13, 2004, which was the day the category 4 hurricane hit the island, had to have seen the spectacular arching tunnel formed by the forty-foot Australian pines over Periwinkle Way. In fact, they had to pass under it if they drove on the island's main thoroughfare.

For me, traveling through that arch, under all those feathery pines, was like entering a magical portal. I think I loved and appreciated that arch and those trees even more than I loved Sanibel's plentitude of shells. Whenever I passed through it, awe and wonderment were stirred in me.

Due to their shallow, intertwined roots, on that fateful day when Charley battered Sanibel with sustained winds of 131 m.p.h., more than a thousand of those beautiful but non-native pines were yanked from the soil, broken into pieces, and downed like a long string of dominos.

As often happens after a hurricane, bright, balmy weather returned quickly, but for five days residents could not return to their homes. The sky may have been clear and the sea may have calmed, but roads were impassable. Debris was everywhere. Virtually every kind of tall tree had been toppled.

When we visited, the roads had been cleared, but a lot of debris still needed to be moved elsewhere. The vegetation that was left standing along Periwinkle and throughout the island was scrubby, squatty, short, and thorny. It looked as if the Sanibel we knew and loved didn't exist anymore. It was painful to see. Like many people, we grieved.

That visit was just for a day. We happened to be on nearby Marco Island for a family event. We were not able to stay on Sanibel for even one night, nor did we want to.

As we went about the island, despite the presence of so much destruction, we witnessed a spirit of fellowship. As often happens in the aftermath of devastation, the best of human nature was drawn from people. At least it appeared that way as we drove around. When we stopped in the grocery store, we witnessed displays of kindness and overheard expressions of compassion, courage, and optimism. We felt strength and hope even more when we talked with a few residents. Nevertheless, we left that day feeling certain we would not return. It seemed as if a chapter of our lives had been completed.

Despite our initial certainty that we would not return, not only did we return multiple times, but over the years since Charley, we introduced many friends and family to Sanibel. We even purchased a little condo on the island.

The first time we returned, we had not known that more than three thousand mahogany, oak, gumbo limbo, and other trees suitable for the climate had been planted. Even though those trees were young and short when we first saw them along Periwinkle, we recognized them as signs of hope and faith. Eventually they would create a new canopy over the roads.

As I remember this about Charley and write about it in 2020, I'm picturing how stripped and messy the island looked after the hurricane, how neat and clean it appeared the first time we returned, and how lush and tropical it looks today.

The takeaway for me is: don't give up. If you value something or someone, even if all seems lost or nearly dead, don't give up. Not on health, dreams, people, or principles. Hold hope, call upon strength, use courage and compassion, and apply love with faith.

And never say "Never."

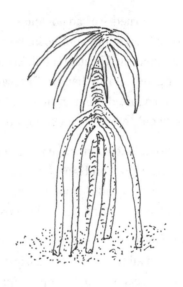

SCREW PINE—*PANDANUS UTILIS*

Roots

To be rooted is perhaps the most important and
least recognized need of the human soul.

—Simone Weil

The screw pine. There are several here. I always notice a
certain one as I round a corner and follow one of the sand-
scattered redbrick paths toward the beach. Without fail, this
tree's distinctive roots grab my attention. Today I wonder.
Why do lengths of roots rise above the soil before meeting
the main trunk? How do exposed roots benefit the tree? Why
is it called a pine if it's not one? The screw part of the name
I get. The above-ground roots are twisted and the branches
are swirly. On this tree, anyway.

The small sign at the base of this tree tells me that its proper
name is *Pandanus*. But with barbed-edged leaves, instead of
screw pine its common name could be "ouch tree." With its
stilt-like roots, it could be "walks-on-stilts tree," or "ready-
to-run tree." I picture this tree breaking loose, taking an
awkward step, and gamboling off. I chuckle and decide I'm
calling it the "root lessons tree" today because its exposed
roots are stirring me to think of how we humans are rooted.

Dictionary definitions of "rooted" include phrases such as "deeply and firmly established," "to be the source or origin of something," and "strong and difficult to destroy." Some say we are becoming less rooted, dangerously so.

I personally appreciate being rooted in a family, in a kinship of friends, and in a community. Yet I'm missing the rootedness I experienced as a child within a large extended family with all members living nearby, many within walking distance. We didn't plan ahead to spend quality time together, we just did it. No one planned to "instruct" or "enrich" us kids. It just happened. Naturally and effortlessly. I learned of perseverance listening to my grandparents' stories of the Old Country. I learned the relationship cementing power of gentle humor witnessing my father and uncles play harmless practical jokes on each other. I learned patience and compassion seeing them care for their shell-shocked brother. In a flash I'm seeing all this and more in my mind. Holiday rituals. Traditions. Knowing there was always someone who would show up and help whenever help was needed. Of course, they all bugged me at times. Each could be quirky. But I felt safe and securely rooted with them.

Like many contemporary families, ours is now spread throughout the country. We're fewer in number now too. The three generations whose members I knew earlier in life have passed. My great-grandparents. My grandparents.

Eleven aunts and uncles. Dozens of great aunts and uncles. My mom and dad.

Now my husband and I, with our few siblings and cousins, are the elders. Conditions are vastly different. Are we providing a strong sense of grounding for our children and grandchildren as our predecessors did? I don't know. I hope so. Will the children learn what we learned by osmosis? Children will always learn vicariously from what is happening around them. They hear what we say. They watch what we do. I hope our words and actions are aligned. Even if they are, the extended family for our grandchildren is small. My husband and I are the only relatives, other than their parents, whom they see frequently. The time we spend with them is potent. We try to give them our full attention. We tell them our stories and the stories we heard from our elders. Mainly, we want them to know they are loved and valued.

Actually, shouldn't all children know this?

I'm sensing that the "root lessons" persona of this tree has more to teach. I'm ready to learn, but not today. Another day I'll ask.

With a sigh, I reach into a pocket and retrieve my cell phone. I've photographed this tree in the past, many times. But I want a photo of it in today's light. I walk around it, step back, and look from different angles. I don't see just the separate roots this time. I look to where they connect at the trunk.

I'm noticing how each exposed root meets the others in the center to form one being. Whole. Strong. Healthy. Looking at this tree differently, I feel different, lighter. I'll take one more photo, then continue down this path.

Click.

I smile. It's a small thing, a smile. It costs nothing. And yet, I suspect smiling has the power to strengthen rootedness, our own and our rootedness with others. Smiling helps us feel connected—be connected—like the roots joining the tree trunk. According to Mother Teresa, "Peace begins with a smile." I get the idea to smile at every person I pass as I walk on.

Home Sweet Home

Give yourself a gift of five minutes of contemplation in
awe of everything you see around you. Go outside and
turn your attention to the many miracles around you.

—**Wayne Dyer**

I'm "sheltering in place" in central Florida in accordance with
steps to stop the spread of the COVID-19 coronavirus. My
husband and I have been at home, almost exclusively indoors,
for three weeks so far. Missing our usual activities, including
spending time with our grandchildren, we're feeling a mix of
sadness, anxiety, and irritability. But we're also doing what we
can to increase our hope and gratitude. Today, I follow Wayne
Dyer's suggestion to spend five minutes outdoors, but I'm in
my own small backyard in Clermont, not on Sanibel Island.

Upon stepping out, I detect the scent of rosemary wafting on
air currents from the large bush growing at the corner of our
house. I plan to cut some for tonight's dinner, but first I walk
straight ahead to the butterfly garden we've been nurturing
just twelve feet from the back of our house. I immediately
notice the impressive overnight growth of the silk floss tree
that we planted last year. For most of the year the tree seemed

not to grow at all. In fact it looked like it wouldn't survive. But since the start of spring a few weeks ago, this thorny, green-barked trunk is growing like the famous beanstalk.

Although I'm appreciating the scent of rosemary, the sight of tree growth, and the blooming perennials, my feelings still seem a bit numbed. I'm not experiencing awe. Nevertheless, being outdoors and close to nature feels like a gift, not quite a miracle, but a welcome gift during this difficult time.

I turn back toward the rosemary bush when a small gray ceramic tile catches my attention. I haven't noticed it for some time, maybe months. This little nondescript tile has moved with us from house to house, across the country, throughout the years. From a distance it blends with its surroundings. It seems to have chameleon qualities; at times it appears gray, then tan, then green, then even pink. I hadn't seen these qualities when I purchased it. I just liked its simplicity. One has to be quite near it to see the word "welcome" molded into the ceramic along with images of a shy hummingbird and a demure blossom.

For the past few years this tile has rested against the base of a tall, mature, live oak. The tile has been a symbol of our welcoming attitude toward the many birds and squirrels that live in the vicinity of our home. Because the sign drew my attention, I approach to take a closer look. I squat close to the ground, trying not to fall headfirst over my bent knees, and I experience unanticipated amusement and appreciation. Across the top edge of the tile, a gecko rests, nonchalantly, above the word "welcome." He seems to feel safe, comfortable, and confident, because when I get closer he remains. For a minute or so I keep still, holding my balance while watching the grip of his feet, feeling wonder at this reptilian life form that resembles the dinosaurs of prehistoric times.

So, after all, in just about five minutes outdoors, I did experience some awe and delight. As I pick a sprig of rosemary, I realize that rather than resist this limbo time between the normal we took for granted and the new normal yet to be discovered, I can appreciate what is, right now.

Forks in the Road

When you leave a beautiful place,
you carry it with you wherever you go.

—Alexandra Stoddard

I'm standing at a place of choice, a point where five garden pathways meet. I can head toward a bridge that would convey me over a natural growth area to the beach. Another leads to a gate which allows entry to a pool. Another runs parallel to the shoreline. One path leads past the ponytail palm where more pathway choices await. The fifth path enters what I call the sanctuary garden, which holds a soothing fountain under a canopy of tree branches, surrounded by a variety of greens.

Before I enter the sanctuary garden, I pause here, at the crossroads, to soak up the beauty and peace of this place. I try to capture a sensory memory to draw upon in the future. In times of stress and uncertainty, I find that it helps me to call to mind places where and moments when I saw beauty and felt peace. I close my eyes and take a deep breath. I smell the air. I wiggle my toes in my shoes and feel the strength of my stance. I feel a slight breeze in my hair. I hear birdsong nearby. I open my eyes and look down each path. I recognize that I have choices, as we always do.

The memory of my experience of this moment is a gift I give myself to open with my imagination in the future, to ground me at some time when I feel confused or unsure which way to go.

Epilogue

Nature is not a place to visit. It is home.

—Gary Snyder

The entire earth is a garden, abundant with lessons from plants, animals, and all of nature to remind us of our innate gifts and how to use them to live with well-being and peace. The majority of experiences presented in this book took place on Sanibel Island, Florida, in the USA. Following are a few additional stories from elsewhere on this beautiful planet.

Do You See What I See?

My destination is no longer a place,
rather a new way of seeing.

—Marcel Proust

For almost ten years, we lived in Sedona, a small city in northern Arizona. We moved there when I was in treatment for lymphoma. It is a powerful place to heal, grow, and learn life lessons. Sedona is popular with artists, spiritual seekers, and outdoor enthusiasts, but it is best known for its red rocks.

Nestled in and surrounded by cliffs, mesas, enormous buttes, spindly spires, and limestone canyons, to me the area looks like a natural, spectacular, colossal rock garden. Juniper, cypress, acacia, cottonwood, mesquite, and ponderosa pine scent the air and soften and complement the towering rocks and limestone walls. Further complementing the rocks is the deep cyan-blue color of the sky.

Besides the intense coloring, Sedona's rocks are well known for their fascinating shapes. A favorite pastime of visitors is to find and identify the named formations. Among the most known are Cathedral Rock, Bell Rock, Merry-Go-Round,

Snoopy Rock, God's Chair, Steamboat Rock, and Courthouse Butte. From a given area, a formation is clearly recognizable as what it is named for. However, seen from a different vantage point, that same formation may seem inaccurately named, appearing nothing like its moniker.

We used to live near what is called Coffee Pot Rock. From that area, the formation truly resembles an old-style percolator. But, when seen from various spots along hiking trails, that same rock can look like a chicken or just a really big rock.

When a group of friends from around the country came to visit, one of the things they wanted to do was hike to several of the famed rocks to see what they looked like up close. Snoopy Rock was their first choice. We hadn't yet taken that trail ourselves, so it would be a discovery for us too. Everywhere along the trail, through turns, up climbs, and down descents, there were countless views in all directions. Of course they changed as we moved; the views are specific to where one stands. When we reached where Snoopy seemed to be from a distance, we were unsure of which rock or rocks made up the formation. From up close, we saw the dusty trail, rocks of all sizes, scrubby desert trees and bushes, but no Snoopy.

God's Chair is another example of something that looks entirely different from near and far. My brotherin-law especially liked this formation. He took many photos of it under a variety of conditions—glowing golden at sunset, dusted with snow, and shrouded in fog. One day he and his wife hiked to it. The trek proved puzzling. They said that what from a distance appears to be an impressive giant chair is actually composed of two enormous but disconnected rock slabs. Yet despite this, from a distance it is appreciated for the symbolism it holds and the feelings it can kindle.

Many people recognize the same clear, beautiful, amazing, and enduring images in the red rocks from a distance. But up close, though interesting and attractive, they usually don't look like anything meaningful or identifiable. Our life situations and problems seem to be similar. Often we are too close to an issue, unable to understand it, find meaning in it, or even clearly name it. But when we step back to observe it from a different perspective, especially if we can do so with faith, receptivity, and even enthusiasm, we can gain clarity, guidance, and inspiration.

A Tree Grows in Sedona

For there is hope for a tree, if it is cut down, that it
will sprout again, and that its shoots will not cease.
Though its root grows old in the earth, and its stump
dies in the ground, yet at the scent of water it will
bud and put forth branches like a young plant.

—Job 14:7-9

Like many Sedona residents and visitors, my next-door
neighbor and friend Joanne felt deep love for nature. She
would call to tell me to look out the window when a mama
quail and her quarter-sized babies were running about
behind our houses and when the javelinas were rooting
around among the scrubby pine trees. She would ask if I
noticed how particularly beautiful the clouds were that day.
Joanne was a gentle, sensitive woman who rescued bugs and
was careful to use only earth-friendly products, well before
it became common.

Joanne was especially fond of a tree that stood in front of her
townhouse. She would often sit on her porch in the morning
sipping tea and in the evening sipping a glass of white wine,
listening to its leaves rustle and watching them move. The

tree shaded the porch and graced her home. The tree was like a friend to her. She spent a lot of time in its company, and, since she lived alone, the tree felt like a companion.

One day Joanne noticed small dark spots on the leaves. Close inspection revealed that millions of tiny bugs had invaded her arboreal friend. It seemed to happen overnight. She had not seen them the day before. She hated to harm the bugs but wanted them to leave her beloved tree. She tried spraying the leaves with a vinegar solution, but she couldn't reach much of the tree anyway. She called the homeowners association to report the problem and asked that someone take a look.

Joanne was horrified when, a few days later, she came home to find that the tree had been chopped down. In fact, by the time she arrived, the cleanup was almost complete. All that was left was a circle of stones around where the tree had been.

She regretted reporting the problem. "I should have tried some things myself, taken some other approach...at least been here when it happened. I could have prevented it," she repeated to herself and to all who would listen.

Day after day, Joanne mourned her missing tree. Where there had been beauty and vibrant life, there was now a barrenness in front of her home—no singing birds, no swaying branches, no soothing shade, no musical rustling of leaves. Just a circle of stones, like a grave marker, around where the tree had stood.

Many months later, a shoot appeared from within that circle of empty red earth. Joanne thought it was some sort of weed. "At least something green is growing there," she thought. Then another shoot appeared, and another. Soon, it looked like a shrub. But in a year's time, it was again a tree…the same tree. For, although all that stood above the ground had been removed, the roots continued to live below ground.

Witnessing this from next door was a gift: watching the tree's resurrection itself and watching Joanne's hope and faith double. She loved the tree even more than before.

The return of Joanne's resilient tree symbolizes the strength and hope that lives in all of us and reminds us to keep courage and have faith. Something new can spring forth from abandoned dreams, stuck relationships, depleted finances, from challenge in any branch of our lives. There is hope for what may appear dead or dying in our lives.

Gregg Buckthorn

When you change the way you look at
things, the things you look at change.

—Max Planck

Not too far from our home in Sedona is St. John Vianney
Church, which boasts lovely grounds. One day after visiting its
gardens, I walked the church's small labyrinth. My thoughts
were of an elder relative, Lucky, who had died recently. He
loved Sedona, especially the gardens of the church and the
church itself. Thoughts of Lucky led me to contemplate and
pray for all of my family and friends who had died.

Before leaving the grounds, I pinched a tiny piece of juniper
and held it near my nose while I walked. Its fragrance always
has a strong immediate effect on me, as many smells do for
many people. Juniper uplifts and energizes me, yet also
calms me, and leaves me feeling as if my physical health is
somehow boosted too. Continuing on, I crossed Soldier Pass
Road toward one of the many trailheads in the area.

I was still thinking of the deceased when I started hiking on the Sunrise Trail. Shortly into the walk, I noticed a sign with the name "Gregg" and "The Buckthorn Family."

Although I had not known him, I wondered about Gregg and the Buckthorn family that had placed this commemorative marker on the trail. Perhaps he was an avid hiker. Maybe he had donated money to maintain the area parks. Perhaps Gregg was a man who showed deep reverence for nature. I said a quick silent prayer for him and his family before moving on and continuing to enjoy the sights and smells of the trail's surrounding flora.

My thoughts moved in a new direction as my gaze lifted beyond the nearby scenery out to the wide red rock vista. As I walked on, I noticed that I could not identify God's Chair or Merry-Go-Round from where I was. But even when I didn't recognize any of the famously named formations, the rock itself was beautiful to me, in fact maybe more so, because I was not seeing it in a predefined way or as a set of caricatures.

After a few more steps, my focus came back to the sights along the trail. I noticed another marker. This one said, "The Cashew Family."

I laughed out loud when I realized that the bronze plates I had seen were not to honor people who had died; they were there to identify plants along the trail!

My recent thoughts of loved ones who had passed had clearly primed me to be thinking through the lens of that topic. So my mistake was understandable. It was also instructive for me as a reminder to think about my thinking. To examine my existing thought patterns. To avoid incorrect interpretations or unhelpful judgments. To try to notice when I am making assumptions about another person's feelings or ideas. To question my own perspective and be willing to see from other perspectives.

Still chuckling, I wondered how often my previous experiences influence my understanding of a situation in a way that leads me to an inaccurate conclusion. Of course we usually have no idea when this happens. At least not at the time; sometimes we'll have the benefit of learning this later. Usually it's not as quickly or as humorously as it was for me with Gregg of the Buckhorn Family.

Animal Wisdom

Animals are not only beholders of great beauty,
but they are also beholders of ancient wisdom.

—Molly Friedenfeld

Nature gives us guiding messages through animals as well as through plants. We have much to discover about intelligence and sensitivities in both the plant world and the animal kingdom. And discoveries are happening. Exciting ones. Especially the ones we have through direct experiences with animals. More will happen as we accept that every living being is sentient to some degree and when we recognize the communication, cooperation, and caring that happens in the natural world.

In recent years we've been hearing new theories and findings about the intelligence and sensitivity of dolphins, whales, elephants, and other animals. In some societies this is not new. In many indigenous cultures, animals have been honored for generations. Among them are the cow, elephant, tiger, monkey, dog, and cat. There are long-standing reverent beliefs that animals perceive reality in ways we don't yet understand and with a depth of wisdom we don't yet appreciate.

In some Native American cultures there is the belief that when an animal shows up in a significant way, frequently or in an unusual setting, either in dreams or waking life, there is a guiding life lesson for us from that animal.

This has happened several times for me. The first animal was the bear. I had many vivid dreams about bears. I saw plush bears in store windows. I kept seeing bears in films and commercials. Bears seemed to be everywhere. It turned out to be a time for me to "hibernate," then to leave a job, rest again, and then fully follow my heart.

There was a period during which snakes were my guide, showing up in a few dreams and several times in real life. That message had to do with shedding old habits, stretching myself, and becoming comfortable in a new skin, a new role.

Another time it was dogs. At the Franciscan Renewal Center in Scottsdale, Arizona (a desert retreat center known as the CASA), several dogs followed me around on the grounds while I was there for a private retreat. They weren't in a pack. It was three different dogs showing up individually at three different times. The intriguing thing is that, other than during the Blessings of the Animals service when animals of all kinds can be seen, from horses to goldfish to dogs to cats to hamsters, I don't recall seeing a dog on the CASA grounds at any other time I was there for a program, either before or since that retreat.

During the retreat, the first dog showed up while I was on a meditative walk and remained near, matching my slow pace for some time before wandering off. Later, another dog followed me at a regular pace, came next to me, and then passed me and went ahead. During another outdoor activity, I sensed something by my feet. I hadn't seen or felt anything approaching, so I was surprised when I looked down and saw a sweet dog sitting there, looking up at me. Once we made eye contact, he stood and walked away. None of these events seem significant on the surface, yet they felt meaningful. I'm inclined to think that many of us have had similar experiences.

Within a week of that retreat, while I was driving along the narrow cliff road in Oak Creek Canyon between Sedona and Flagstaff, I saw a large dog on the side of the road along a stretch where there are no facilities or structures, nor a place to pull off the road. What's that dog doing here? I thought. It was unsafe for the dog to be there and potentially unsafe for me to stop in that area. Fortunately there was no traffic on the winding road at that time. After positioning the car near the dog, I stopped, not knowing what I would do, but knowing I would do something. Seeing the dog close up, I could tell I would be safe to approach him. At least that's what I thought because he looked healthy and well treated. When

I opened the door, the dog jumped in, hopped over the front seat to the back seat, and sat straight up, acting as if this was something we did routinely. Even though it was at first an odd feeling to have this large unknown dog in the car, it soon felt as if he were a loyal friend to me and that he expected loyal friendship from me as well. There was a sense of mutual trust. He sat close with his face quite near my shoulder. Whenever I glanced back at him, he seemed confident that either he or I knew what we were doing and where we were going. By the time I got to Flagstaff, I loved that dog. I almost wanted to keep him. Of course, I couldn't. I believed he was very much loved by someone or someones, perhaps a family. He didn't have a collar, so I brought him to a no-kill shelter and felt quite sad to say goodbye to him. The attendant assured me that he was clearly a healthy, smart, well cared for dog, and that his person was probably searching for him. Only hours later I was happy to learn that he had been reunited with his family. I never learned the details of how he got lost, but I felt that my finding him was helpful for me finding *me* at the time.

Dogs are symbols of loyalty and reminders to be true to oneself. The experiences with these four dogs, especially with my car companion, prompted me to make several life changes that required self-trust and increased authenticity. It's many years later now, but when I picture those soft brown eyes, I'm reminded again to be fully loyal to myself.

At another time, my messenger animals were turtles. My peak turtle encounter happened on the island of St. John in the US Virgin Islands. Frank and I were fortunate to finally vacation on that simple, unassuming island in 2017. It had been on my "someday" wish list since 1977. We stayed at Caneel Bay, a low-key resort within the USVI National Park, on the grounds of an old sugar plantation where wild donkeys roamed.

Both plants and animals contributed to the earthy serenity of the island and especially where we stayed. Every day we saw countless donkeys and deer and spotted a mongoose now and then. Whenever and wherever we encountered the animals, we felt safe sharing the grounds and the beaches with them. The comfortable ease between us and the animals felt natural and healthy, like a peaceful paradise.

But it was the sea turtles that had a lasting impact on me, one in particular. Four of the seven species of sea turtles in the world are present in the waters around St. John— hawksbills, greens, leatherbacks, and loggerheads. Every day we saw many.

What I know about them moves me to awe and reverence. Sea turtles have the ability to migrate hundreds, sometimes thousands, of miles from their feeding grounds to their nesting beachs. Adult females return to nest on the very same beach where they hatched. How they do that is so remarkable that it can be hard to fathom. Somehow they use the earth's magnetic fields to navigate.

Having such high regard for sea turtles, every time we saw their heads pop up from below the surface to take a breath, we felt a thrill. Each sighting stirred gratitude and more awe.

I actually cried out while snorkeling when a giant turtle came into my view, swam to a spot directly under me, and began to eat. Being able to watch her nibble on seagrasses felt like a privilege. And she seemed to be smiling. I was mesmerized. After a while, in her own time, the turtle swam upward and surfaced for air right beside me. With her so near, I felt even more gratitude and love for turtles, especially that particular one. Being near a large creature whose body and experiences of living are so different from mine was a gift.

I've learned that the turtle is a sacred figure in Native American mythology. It represents Mother Earth and symbolizes good health and long life. And the turtle encourages us to stay on our paths toward peace with perseverance.

Shortly after the encounter, I noticed that my feelings toward that turtle were motherly. I realized I'm capable of caring for all creatures with a deep, protective love, a nurturing, maternal love. We're all capable of such love for all beings.

Afterword

Dear Reader,

Thank you for reading my book. I hope that during your reading, some of your own garden and nature memories surfaced, bringing happiness into your day with fresh new insights for living your best life now. I hope the stories stirred your dreams, reminded you of your valuable innate gifts, and prompted ideas to enhance your peace and well-being.

I encourage you to further explore your Twelve Gifts of strength, beauty, courage, compassion, hope, joy, talent, imagination, reverence, wisdom, love, and faith. They will help you to grow and thrive.

Also, please visit my website, CharleneCostanzo.com. Among other resources, there you'll find an abundance of photographs to help you visualize yourself in the garden. You may also wish to sign up for my daily inspirational email *Today's Touchstone*. I always welcome hearing from readers. Look for me on Facebook. Please consider posting a review on Goodreads, Amazon, and other online bookstores.

Write to me through my website with comments and your own stories and life lessons from nature.

Wishing you the best of life's gifts,

—CHARLENE

For Reflection and Discussion

Several times in *The Twelve Gifts from the Garden*, Charlene mentions that a particular experience has become a touchstone for her, a reminder of a particular principle to guide her in navigating life's challenges. What are some of your own memories that serve as touchstones for you?

Consider your own encounters with plants, animals, the environment, the weather—any part of nature. How did each of those experiences demonstrate one or more of the Twelve Gifts to you? What might you do to further expand your awareness of your innate Twelve Gifts while in nature? How might you use this increased awareness of your gifts to improve your well-being and sense of peace?

Bridges provide the means for us to get to where we want or need to go, conveying us over rough or dangerous areas. Bridges were present in several stories, characterized by feelings varying from joyful anticipation to pleasure to hesitation to trepidation. What thoughts and feelings do these different bridges and crossings stir in you? Bridges also represent connections. What connections or reconnections might you like to make in your life?

Finding the beach garden made of flotsam and jetsam was a surprise in and of itself to Charlene. Hearing the story behind it astonished her. What was your experience in reading this story? What was your takeaway?

From early childhood to the present day, Charlene has used imagination for many purposes, from simple playfulness to helping herself heal. How did her imaginative treatment of trees as having personalities affect you? How do you imagine that plants experience life? Consider how you used imagination as a child and how you use it now. How has it best served you? Consider dreams, goals, creative expression, coping, playfulness, and interacting with other people, perhaps people you find difficult. How might you engage your imagination more, should you wish to?

"The Silk Floss Tree" is about coming to terms with thorns present in plants, but more importantly in the personalities of people with whom we interact. What did this story bring up for you? Who or what in your life might you choose to see differently?

Through the story "The Great Pretender," Charlene invites you to ask yourself the following questions: Have you ever acted as if you were familiar with something, or someone, or a topic being discussed, when you really weren't? Perhaps everyone in the conversation chimed in with knowledge about the topic. What did you do? Act as if you knew about it in an effort to fit in? Remain silent and hope that no one would ask your opinion? Or did you courageously and unapologetically admit your unknowing and ask for information? How have you felt when it took courage to be honest, real, and vulnerable in a situation? How do you feel when you say, "I don't know?"

Consider some of life's storms you have weathered, either alone or with others. Which of your innate Twelve Gifts did you draw on? How did you survive or triumph? In what ways have you been strengthened by your adversities? What did you learn? What parts of your life may have actually improved as a result of those storms?

"Down but Not Out" is the story of the large tree lying on the ground, with most of its roots pulled up, but still living and growing. What was your experience in reading this? What was your takeaway from the story of this tree?

"Photo of the Day" discusses celebrating ordinary days and moments in addition to what we normally think of as extraordinary and special ones. In what ways, if any, do you do this? Might you do something differently to honor and cherish the gift of each day?

Throughout several stories, Charlene presents ideas about the way people treat the environment. These stories offer much to reflect upon, write about, and discuss. They also raise issues that are highly sensitive to some people and about which there are different perspectives. Consider your own beliefs and ideas about human interaction with insects, plants, animals, and resources. What are your thoughts about the contrasting needs of people and the other living things on earth? And how might we listen to understand all perspectives about a given issue?

The theme of beauty is present in many stories in this book. What did these stories stir in you? To what extent are you comfortable with your own beauty and with discussing it? What thoughts and feelings arose as you read "Upside Down"? How are you feeling about your inner beauty right now? In what ways might you view beauty differently?

Joy is found where there is gratitude. "Joie de Vivre" presents feelings of joy in the beauty of nature, fine weather, the companionship of people and animals, and lightheartedness. What does joy mean to you? What stirs joy in you?

What does reverence mean to you? How do you experience reverence? Which stories best represented and expressed reverence from your perspective? In what way might you perceive reverence differently after reading these stories?

Weight, appearance, and body image are issues, in some form, for many of us. What was your experience reading "Scrambled Eggs with a Side of Compassion"? Which of the Twelve Gifts felt present for you?

In several stories, death is faced in different forms—trees appear dead, illnesses are faced, loved ones pass away, and the deceased are prayed for. What feeling is most present for you right now regarding death? In what ways and places did you experience hope in these stories? Faith? Courage? Other gifts?

In several stories, including "Time Travel" and "Family Tree," something in the garden prompts Charlene to remember and reflect on childhood events and interactions with family members. Recall time you have spent with relatives in the past, particularly elders. How did those times shape you, your outlook, and your beliefs? What qualities and interests did they nurture in you? Which of the Twelve Gifts did you witness in relatives? Consider what you'd like your young loved ones to know about their own inner gifts.

Think about, write about, or discuss your most loving experience with an animal. When have you encountered an animal in a profound way? A humorous way? What lessons have animals provided for you? What is your understanding of animal sentience and consciousness?

Charlene discusses a change in perspective in several stories, particularly those set in Sedona, AZ. She believes that seeing differently can help bring peace within and among us. Recall one or more times when you deliberately shifted your viewpoint. What brought this about? How did it affect you?

What changed as a result of this shift? To further challenge yourself, consider choosing a topic about which you feel strongly. Develop a considered, logical argument against your normal viewpoint, as if in a debate. What might you learn about the perspectives of people on the opposite side of the position? How might this help you better understand your own ideas and beliefs? How might it help you tolerate, respect, or even appreciate those with whom you currently disagree?

Acknowledgements

The Twelve Gifts from the Garden started long ago as a single seed of inspiration and grew slowly over time. Thank you, family and friends, who in your own special ways helped me nurture it into this published book. You may have listened, read, encouraged, criticized constructively, or prayed. I also appreciate your patience, kindness, and understanding when I needed to be in my own little writing world and was unavailable. Please forgive me if your name doesn't appear here but should.

Kudos and credit go to illustrator, Mary Lou Peters, to designer and creative consultant, Karen Heard, and to my agent, Gary Krebs. Gratitude goes with cheers for all at Mango Publishing. I appreciate your vision and enthusiasm for *The Twelve Gifts from the Garden*. Thanks go to Brenda Knight, Robin Miller, Yaddyra Peralta, Lisa McGuiness, Morgane Leoni, the sales reps, and everyone else on the Mango team. Thank you all, for giving your all.

Special thanks to Reeve Lindbergh, Adrienne Falzon, Angela Howell, Gloria Gaynor, Gina La Benz, Charles Sobezak, and Maryann Ridini Spencer.

More thanksgiving goes to Alicia, Anita, Angie, Annmarie, April, Barb, Barbara, Bob, Carol, Sister Caroline, Cheryl, Diana, Donna, Glenice, Joanie, Jan, Jan, Jan, Jill, Jim, Kari, Kathy, Keith, Krista, Kyra, Linda, Linda, Linda, Marilyn, Margaret, Maria, Mary Lu, Mary Margaret, Mary Margaret, Mike, Monica, Nancy, Peg, Peggy, Rebecca, Ruth, Sandy, Steve, Susan, Terri, Tina, Tony, Tricia, Veronica, Zoe, and the women at the Women's Prosperity Network.

Finally, I must give special recognition to my husband, Frank, and our daughter, Stephanie. Frank, thank you for your patience, technical assistance, and general support. I could not have done this without you. Stephanie, your substantive editing was stellar. I could not have done this without you either. Working with you is a joy.

With gratitude and love to all.

Beyond the Botanical Garden

Every person who crosses the causeway to Sanibel is greeted by the sign, "Welcome to Our Sanctuary Island: Do Enjoy. Don't Destroy."

I must acknowledge those who first imagined Sanibel as a sanctuary island and all who uphold the vision of Sanibel as a place where nature is enjoyed, respected, and preserved.

Most of the stories in this book were inspired at Sanibel Moorings resort, which states its core purpose is "to provide memorable experiences in harmony with nature." Thanks to the mindful management, the staff, and the board leadership over the years, my experiences have been memorable and in harmony with nature as well as renewing for my body and enriching for my soul. Had they not, this book would not have been written, at least not this way. To everyone on all the teams who polished this gem since 1974, keep it sparkling now, and will in the future: Thank you. I especially acknowledge the gardeners, all who brought their own special visions and gave loving care through the decades, back to the first, who was an avid botanist, and onward to all who will continue to care for these six lush acres of earth.

And, in addition to my experience at the Moorings, the vitality of nature throughout the island nurtured my vision and enthusiasm for this book. Therefore, with gratitude I acknowledge all who protect Sanibel and promote nature-supporting choices, including the City of Sanibel, Sanibel-Captiva Conservation Foundation, J.N. "Ding" Darling National Wildlife Refuge, Bailey Homestead Preserve, Clinic for the Rehabilitation of Wildlife, Sanibel Chamber of Commerce, Sanibel Communities for Clean Water, Native Nursery at Bailey's Homestead, Bailey Tract, Frannie's Preserve, Pond Apple Trail, Bailey-Matthews Shell Museum, Sanibel Historical Museum and Village, Calusa Shell Mound Trail, the Community House, and the caring and conscientious residents of Sanibel and Captiva.

Enter the Garden

So that readers can share, virtually, in the experience of Sanibel as a sanctuary island and see the botanicals and the pathways described in this book, photographs are available on my website, www.CharleneCostanzo.com. If you're wondering what the pom-pom palm and the passion flower look like, if you'd like to see Sea Legs and Smiley the spider, if you would like to know more about living the Twelve Gifts in harmony with nature, please visit my website and explore. I hope you'll visit often. Come for a virtual vacation, a healthy getaway, inspiration, and to simply enjoy the plants featured in *The Twelve Gifts from the Garden: Life Lessons for Peace and Well-Being*.

About the Author

As well as being an award-winning author, Charlene Gorda Costanzo is a speaker, workshop facilitator, wife, mother of two adult daughters, and grandmother of twins. She holds a BA in philosophy from St. Bonaventure University and an MA in spiritual psychology from the University of Santa Monica.

Originally from New Jersey, she has resided in New York, Texas, Arizona, and Florida. During a one-year book tour to launch *The Twelve Gifts of Birth*, Charlene and her husband enjoyed living in an RV in forty-eight of the fifty states.

The Twelve Gifts collection of books began in 1987 when Charlene wrote *The Twelve Gifts of Birth* as a life message for her own (then teenage) daughters. Twelve years later she published the book and then spent a year sharing and discussing its message in schools, shelters, prisons, churches, and hospitals throughout the United States.

The Twelve Gifts for Healing was written while Charlene was in treatment for advanced non-Hodgkin's lymphoma in 2001. "Cancer led me to examine my convictions and look at these life gifts more deeply. Truly, they helped me heal," she says.

The Twelve Gifts in Marriage emerged from the ups and downs, ebbs and flows, and hurts and healings that are a part of every long-term marriage or committed relationship.

The Thirteenth Gift is a novella that celebrates the wonderment that dwells in our hearts and helps us to see the world with reverence, gratitude, and joy.

Touchstones: Stories for Living the Twelve Gifts offers true tales that demonstrate life's gifts at work in everyday life.

About the Illustrator

The spot illustrations in *The Twelve Gifts from the Garden* were rendered by Mary Lou Peters, a Michigan artist who has been drawing and painting for thirty years. Mary Lou's pen and ink drawings convey the tropical feeling of Sanibel Island, where she often visits and offers classes. Her goal is to capture the fun nature of wherever she is by using a whimsical approach to common themes. She hopes that her drawings leave readers smiling and wishing they were in the gardens she has drawn.

Mary Lou's artwork is in collections worldwide and may be viewed on her website, www.maryloupeters.com and on her Facebook page, Watercolor Paintings by Mary Lou Peters. Mary Lou has illustrated three other books, *Mackinac Island Meditations*, *Garden Meditations*, and *Jerry's Etiquette on the Go*.

Mango Publishing, established in 2014, publishes an eclectic list of books by diverse authors—both new and established voices—on topics ranging from business, personal growth, women's empowerment, LGBTQ studies, health, and spirituality to history, popular culture, time management, decluttering, lifestyle, mental wellness, aging, and sustainable living. We were recently named 2019 *and* 2020's #1 fastest growing independent publisher by *Publishers Weekly*. Our success is driven by our main goal, which is to publish high-quality books that will entertain readers as well as make a positive difference in their lives.

Our readers are our most important resource; we value your input, suggestions, and ideas. We'd love to hear from you—after all, we are publishing books for you!

Please stay in touch with us and follow us at:

Facebook: Mango Publishing
Twitter: @MangoPublishing
Instagram: @MangoPublishing
LinkedIn: Mango Publishing
Pinterest: Mango Publishing

Newsletter: mangopublishinggroup.com/newsletter

Join us on Mango's journey to reinvent publishing, one book at a time.

CPSIA information can be obtained
at www.ICGtesting.com
Printed in the USA
JSHW051649091020
8631JS00004B/11